THE LIVING RECORD OF SCIENTIFIC HISTORY

Conversations with CN Yang

THE LIVING RECORD OF SCIENTIFIC HISTORY

Conversations with CN Yang

Lizhen Ji

University of Michigan, USA

Liping Wang

Translators
**Yi Song, Menghan Li,
Huiyan Ding & Neta M. Cahill**

Published by

World Scientific Publishing Co. Pte. Ltd.

5 Toh Tuck Link, Singapore 596224

USA office: 27 Warren Street, Suite 401-402, Hackensack, NJ 07601

UK office: 57 Shelton Street, Covent Garden, London WC2H 9HE

Library of Congress Control Number: 2023945968

British Library Cataloguing-in-Publication Data
A catalogue record for this book is available from the British Library.

B&R Book Program

《百年科学往事——杨振宁访谈录》
季理真、王丽萍 编著
Copyright © 2021 by East China Normal University Press Ltd.

THE LIVING RECORD OF SCIENTIFIC HISTORY
Conversations with CN Yang

Copyright © 2025 by World Scientific Publishing Co. Pte. Ltd.

ISBN 978-981-128-493-9 (hardcover)
ISBN 978-981-128-494-6 (ebook for institutions)
ISBN 978-981-128-495-3 (ebook for individuals)

For any available supplementary material, please visit
https://www.worldscientific.com/worldscibooks/10.1142/13641#t=suppl

Desk Editor: Carmen Teo Bin Jie

Introduction to Yang Chen-Ning

Yang Chen-Ning was born in Hefei, Anhui, on October 1, 1922. His father, Yang Wu-Chih, received a Ph.D. in mathematics from the University of Chicago and studied with Dr. Leonard Eugene Dickson. His paper "Various Generalizations of Waring's Problem" inspired Hua Loo-Keng's initial achievements in analytic number theory and led him to a brilliant career in mathematics.

In the summer of 1938, Yang Chen-Ning participated in a unified admission examination as a high school junior and was admitted by the National Southwest Associated University for outstanding academic performance.

In 1942, he graduated from the National Southwest Associated University, and his undergraduate thesis was supervised by Professor Wu Ta-You of Peking University. Later, he was admitted to study in a graduate program at the Department of Physics, Institute of Science (Institute of Physics), Tsinghua University.

In 1944, he graduated from the National Southwest Associated University, and his master thesis was supervised by Professor Wang Zhuxi of Tsinghua University.

In 1945, he won the Boxer Indemnity Scholarship and went to the University of Chicago.

In 1948, he received his Ph.D. from the University of Chicago. His dissertation was supervised by Professor Edward Teller. He spent the next year working as Enrico Fermi's assistant at the University of Chicago.

In 1949, he joined the Institute for Advanced Study at Princeton for postdoctoral studies and began to work with Lee Tsung-Dao. In the same year, in collaboration with Enrico Fermi, he proposed the first composite model of elementary particles.

In 1952, he was granted tenure by the Institute for Advanced Study at Princeton and became a full professor in 1955. He held this position until 1965.

In 1954, at Brookhaven National Laboratory, he and Robert

Laurence Mills proposed a theoretical structure called Non-Abelian Gauge Field Theory.

In 1956, in collaboration with Lee Tsung-Dao, he published a paper that overturned one of the central messages of physics: parity conservation, in which elementary particles and their mirror images behave exactly the same.

In 1957, he shared the Nobel Prize in Physics with Lee Tsung-Dao for work on the so-called parity non-conservation laws. He and Lee were the first Chinese to be awarded a Nobel Prize.

In 1958, he was elected an academician at Academia Sinica in Chinese Taiwan.

In 1964, he became an American citizen.

In 1965, he was elected an academician at the National Academy of Sciences in the USA.

Since 1965, he has been the Einstein Professor of Theoretical Physics at Stony Brook University, New York. The Institute of Theoretical Physics was established, and he served as its director.

In the summer of 1971, he was the first renowned American scholar to visit the People's Republic of China.

In 1977, he founded the National Association of Chinese-Americans (NACA) in Boston with Liang Enzuo and others to promote Sino-U.S. relations.

In 1980, he won the Rumford Prize, one of the oldest scientific awards in the U.S.

In 1982, he became an honorary professor of physics at the Chinese University of Hong Kong.

In 1986, he was bestowed the National Medal of Science by the President of the United States. In the same year, he attended the Conference of Academicians at Academia Sinica in Chinese Taiwan. Also in 1986, he started serving as a Distinguished Professor-at-Large at the Chinese University of Hong Kong.

In 1993, he was elected a fellow of the Royal Society. In the same year, he was awarded the Benjamin Franklin Medal.

In 1994, he was elected a foreign academician of the Chinese Academy of Sciences.

In 1994, he was awarded the Bower Award and Prize for Achievement in Science by the Franklin Institute in Philadelphia, USA.

In 1994, he founded the Institute of Mathematical Sciences at the Chinese University of Hong Kong with Professor Yau Shing-Tung, the first Chinese winner of the Fields Medal.

In 1995, he became an honorary professor of Huaqiao University.

In 1996, he received honorary doctorate degrees from Tsinghua University and Shanghai Jiao Tong University. In the same year, he was awarded the N. Bogoliubov Prize.

In 1997, he founded the Advanced Research Center at Tsinghua University. In the same year, he received an honorary doctorate in science from the Chinese University of Hong Kong.

In May 1997, the International Astronomical Union Minor Planet Center officially named the minor planet numbered 3421 "Yang Chen-Ning Star", as nominated and requested by the Purple Mountain Observatory of Chinese Academy of Sciences—which discovered this planet back in November 26, 1975.

In May 1999, he officially retired, and the State University of New York at Stony Brook renamed the Institute of Theoretical Physics as the "C. N. Yang Institute for Theoretical Physics" and awarded him a first-class honorary doctorate. In the same year, he was awarded the Lars Onsager Prize.

In 2001, he was awarded the King Faisal Prize.

In 2002, he was chairman of the Selection Committee for the Shaw Prize.

At the end of 2003, he returned to Beijing and settled down.

In November 2004, he was appointed as Distinguished University Professor at Hainan University.

In 2009, he served as honorary president of Guangdong Dongguan University of Technology.

In March 2015, he was awarded both an honorary doctorate in science by Taiwan University, as well as the 2014 Honorary Degree by the University of Macau.

In February 2017, he gave up his U.S. citizenship, became a Chinese citizen, and was officially elected an academician of the Chinese Academy of Sciences.

In 2019, he won the Qui Shi Lifetime Achievement Award.

Contents

Preface

There is a Chinese proverb: life rarely lives to 70. It may not be so relevant these days, but it's still not a common thing to live to a ripe old age, especially when one can still remain wise and alert, talk for hours, share one's reflections on life and science, and display great insight and wisdom. For most people, learning new things at a young age is a matter of either duty or interest. Attaining such an old age is unusual in itself, but it is even more unusual for such an old man to have made such a huge contribution to science, and especially to our understanding of the structure of the universe. In addition, despite his worldwide fame and staggering achievements, this gracious man is always kind and willing to talk to young people and listen to their ideas. Is it easy to find such people? Finding someone like this is rare and precious indeed. In fact, such people are particularly rare in the history of science. Although Newton lived to 84, he left science early. Other great figures in the history of science left us even earlier, for example, Einstein died at 76 and Gauss at 77. We are very honored to meet such a person and to have had the great opportunity to engage in in-depth conversations with the great and highly revered physicist Yang Chen-Ning.

We have met with Dr. Yang every year since 2016, so we have had the opportunity to listen to him talk and to engage in in-depth communication with him. The topics are diverse, ranging from mathematics to physics and the social sciences, from mathematicians to physicists, from their great work to their romantic lives, from happy achievements to sad events and tragedies, from academic collaboration to mutual discord. No topic was off the table. Due to various reasons, a few topics and opinions are not suitable for this book, but may be made public in due course. Through this series of interviews, we learned many stories about science and gained a more comprehensive understanding of scientists: how they worked and how they lived; despite their great achievements, these brilliant minds were also human beings—how they behaved like ordinary people, and how unique they were. Years of interviews with Dr. Yang naturally led to a simple question: what is youth? The following poem by Samuel Ullman seems to provide the best answer, as well as the perfect portrait of Dr. Yang's life:

> *Youth is not a time of life; it is a state of mind; it is not a matter of rosy cheeks, red lips and supple knees; it is a matter of the will, a quality of the imagination, a vigor of the emotions; it is the freshness of the deep springs of life.*
>
> *Youth means a temperamental predominance of courage over timidity, of the appetite for adventure over the love of ease. This often exists in a man of 60 more than a boy of 20. Nobody grows old merely by a number of years. We grow old by deserting our ideals.*
>
> *Years may wrinkle the skin, but to give up enthusiasm wrinkles the soul. Worry, fear, self-distrust bows the heart and turns the spirits back to dust. Whether 60 or 16, there is in every human being's heart the lure of wonder, the unfailing childlike appetite of what's next and the joy of the game of living. In the center of your heart and my heart there is a wireless station: so long as it receives messages of beauty, hope, cheer, courage and power from men and from the Infinite, so long are you young.*
>
> *When the aerials are down, and your spirit is covered with snows of cynicism and the ice of pessimism, then you have grown old, even at 20, but as long as your aerials are up, to catch waves of optimism, there is hope you may die young at 80.*

There is, of course, one thing to emphasize about Dr. Yang's achievements. Undoubtedly, his most important contribution was the Yang–Mills Gauge Theory, which was the absolute cornerstone of the fundamental theory of the Standard Model of particles physics. However, people may not have realized that the Yang–Mills theory also inspired the most famous works of Fields Medal winner Simon Donaldson, Abel Prize winner Karen Uhlenbeck, and countless other people.

This book records eight interviews with Dr. Yang from 2016 to 2019. In these eight interviews, the development of mathematics and physics around the world and their mutual influence in the past decades are presented through Dr. Yang's unique perspective. Although

Dr. Yang is a physicist, he has mentioned his mathematician father on many occasions, saying that he learned a lot of mathematics from his father, which was of great help to his academics. In China, the field of mathematics and the field of physics are one big and lively family, including successful mathematicians and physicists such as Yang Chen-Ning, Lee Tsung-Dao, Chern Shiing-Shen, Hua Loo-Keng, Chou Weiliang, Yau Shing-Tung and Tian Gang. There were also younger scholars such as Zhang Shoucheng, Zhang Wei, Yun Zhiwei and Xu Chenyang, who have also done outstanding work, and the influence the older generation has had on them is obvious. This fruitful and hardworking family has had a profound impact on the development of science in China. Of course, it's also an interesting family with its own heritage and stories. It is fair to say that this family represents true Chinese characteristics.

We had hoped that our interviews with Dr. Yang could continue. However, they were interrupted by the COVID-19 pandemic in 2020. October 1, 2021 was Dr. Yang's 100th birthday and we present this book to him as a birthday gift. We wish him good health and eternal happiness. At the same time, we hope that the global pandemic will end as soon as possible so that our interviews can continue into the future, and we can learn even more from Dr. Yang.

In the preface of *History as a Mirror*, Emperor Shenzong of Song Dynasty, also known as Zhao Xu, wrote that a preeminent and honorable man knows the past to refine his virtue, and thus he lives each day with vitality and honesty. That is to say, those with outstanding talents and virtues are usually familiar with past stories, by which they hone their morals, and become as calm as the sky and as stable as a mountain. This way, they are always vigorous and make progress in their morals, literature and art. Naturally, we hope that readers can feel Dr. Yang's enthusiasm for science and life, and learn and appreciate the wisdom he has accumulated. Through his extraordinarily long life, he learned about mathematics, physics and their interactions in the universe from an unusual perspective. We also believe that this book will provide a unique way to understand this fascinating man and touch on the mind of this brilliant intellectual from a new perspective. Most importantly, we hope that readers can enjoy this book by immersing themselves in a pragmatic conversation with a great scientist and wise elder.

It was a daunting task to record all the conversations and transcribe them, especially when these conversations were dotted with foreign names and scientific jargon. First of all, we would like to thank the editor Peng Cheng from East China Normal University Press for her assiduous work. As many famous scientists were mentioned in the interviews, the editors, Peng Cheng, Tang Ming and Zhu Huahua painstakingly prepared an annotated list of names in order to let readers know more about them; here we express our gratitude to them. We would also like to thank Wu Xiang, former deputy editor-in-chief of Higher Education Press, for his meticulous editing work and all the critical support that made this book possible. In addition, we would like to thank Dr. Yang and his secretary, Xu Chen, for the carefully selected and precious photographs. Xu Chen also helped us contact Gao Yuan, the photographer of the book's cover image, and we are grateful to Mr. Gao for his authorization. Naturally, we would also like to thank Wang Yan, president of East China Normal University Press. Without her gracious support this book would not have been published in time.

Finally, as the editors of this book, we must mention that due to the length of the eight interviews, Dr. Yang has not had the opportunity to read the book in detail. We are responsible for any errors and misunderstandings that may arise due to the transcription of informal oral conversation into written language.

Ji Lizhen, Wang Liping
March 3, 2021

The First Interview

Time: August 4, 2016
Place: Institute for Advanced Study at Tsinghua University
Interviewer: Ji Lizhen
Recorder: Wang Liping
Collator: Lin Kailiang, Wang Liping

On August 4, 2016, the editors of this book, Ji Lizhen, Wang Liping and Lin Kailiang, interviewed Dr. Yang at the Institute for Advanced Study, Tsinghua University. For decades, Dr. Yang has made great achievements in the intersection of physics and mathematics, so he personally knows or is familiar with many of the world's top mathematicians and physicists. One of the main purposes of our interview was to learn about these mathematicians and physicists and their stories through Dr. Yang. In addition, Ji Lizhen had spent a year at the Institute for Advanced Study at Princeton and knew mathematicians like Armand Borel, who also had connections with Dr. Yang. Professor Ji wanted to know the story of Yang's relationship with these esteemed mathematicians. During the interview, we marveled at Dr. Yang's amazing memory and energy. The interview lasted two hours without a break and was very cordial and informative. The following is the actual transcript of that interview.

Best Days at the Department of Physics, University of Michigan

Yang Chen-Ning: You said that you studied at the University of Michigan, so do you know George Uhlenbeck[1] and Samuel Goudsmit?

Ji Lizhen: No, I don't. I'd love to hear about them.

Yang Chen-Ning: George Uhlenbeck and Samuel Goudsmit were Dutch physicists who were Ph.D. students of Paul Ehrenfest. At that time, there were more college jobs in the United States, so they went there to find jobs and stayed on after they got their degrees. Their paper, published in 1925, was the first to talk about the 1/2 spin of an electron. There were no people in the United States who had studied quantum mechanics at that time, so they introduced new concepts to the U.S. In particular, they held a summer school every summer for many years, which had a decisive influence on the spread of quantum mechanics to the United States. Some professors were very famous; for example, Enrico Fermi was invited to speak one summer in the 1940s.[2] It was a very important talk, and my understanding of quantum electrodynamics was closely related to his article,[3] which was the focus of his talk that summer. Then, a few years later, they invited Fermi to deliver a speech again, this time on the Theory of Beta Decay. It was also very important at that time, and the Theory of Beta Decay started to take root from that point on. As part of this summer series, they invited Julian Schwinger to talk about renormalization in 1948. Schwinger was a Harvard professor who had later won the Nobel Prize. Although he was still young in 1948, he was widely regarded as the most important young theoretical physicist of his time, so I went there to listen to his speech. Freeman Dyson, whom I knew, was also

---

[1]George Uhlenbeck once supervised two Chinese students, the famous female physicists Wang Mingzhen (1906–2010) and Wang Chengshu (1912–1994).

[2]Fermi actually made his first visit to the United States in 1930, when he went to Michigan to visit George Uhlenbeck.

[3]E. Fermi (1932). Quantum Theory of Radiation. *Reviews of Modern Physics* 4: 87–132. Yang Chen-Ning also mentioned this story in a section on Fermi in *The Story of Several Physicists* (included in *Anthology of Yang Chen-Ning* and *Yang Chen-Ning's Lectures*).

on-site. After he understood Schwinger's method, he went back and wrote an important paper linking together Schwinger and Richard Feynman's work on renormalization within six months. In the 1950s, the influence of the Michigan Physics Summer School waned, in that there were a lot of people from all over the country doing this type of work. It ended in 1954, the year when I was the keynote speaker.[4]

Ji Lizhen: You delivered the last lecture.

Yang Chen-Ning: After I spoke, they closed this summer school—partly because Uhlenbeck and Goudsmit were getting on in years, so they couldn't do it anymore.

Ji Lizhen: Oh, they didn't have that much energy. I didn't know our physics department had such a glorious tradition. Nobody even told me about this.

Yang Chen-Ning: Perhaps you don't have much contact with people in the Physics Department.

Ji Lizhen: We have little contact.

Yang Chen-Ning: When you were there, you knew that the most important man in the Physics Department of the University of Michigan at that time was called Martinus J. G. Veltman. He won a Nobel Prize later.

Ji Lizhen: Yes. A few years ago, his portrait hung outside their building, and then it was removed.

Yang Chen-Ning: He retired and went back to the Netherlands.

Ji Lizhen: Yes. When he won the Nobel Prize, his portrait was displayed outside for a while, but it was taken down after about two

---

[4]The topic was "Introduction to High Energy Physics," which was Yang Chen-Ning's paper.

years. Wasn't there a fellow in Michigan who studied string theory in five dimensions? Klein or something? I can't remember.

Yang Chen-Ning: I'm not familiar with string theorists. They appeared later.

Ji Lizhen: He was in the early stages, and he was the first to put forward five dimensions, but I've forgotten his name.

Lin Kailiang: Oskar Klein.

Ji Lizhen: Yes, Oskar Klein, I have a recollection. I suppose they talked about him before. I'm not familiar with physics at all, and I don't study string theory. Is he in Michigan?

Yang Chen-Ning: Oskar Klein was a very old man, and sadly, he has died.

Ji Lizhen: Yes, a person from a long, long time ago.

Yang Chen-Ning: He was Swedish and was Niels Bohr's assistant. I met him before, and he was probably at least 20 years older than me.[5] He came between 1949 and 1959 when I was at the Institute for Advanced Study at Princeton, so I knew him. I think he was the most important theoretical physicist in Sweden at that time.

Ji Lizhen: Then I suppose I'm wrong,[6] because I am not familiar with physics.

---

[5]Oskar Klein (1894–1977) was 28 years older than Yang. Yang was the recipient of the inaugural Oscar Klein Medal in 1988 and delivered the Oskar Klein Memorial Lecture. In addition, in 1957, when Yang and Lee Tsung-Dao received the Nobel Prize for Physics, Oskar Klein read the award speech at the ceremony. See http://www.nobelprize.org/nobel-prizes/physics/laureates/1957/press.html.

[6]In fact, Oskar Klein was a professor at the University of Michigan in 1923–1925.

String Theory

Yang Chen-Ning: I understand why you asked me this, though. People who study string theory need to think about high-dimensional space.

Ji Lizhen: Yes. And Oskar Klein was the first to come up with this idea.

Yang Chen-Ning: There was another mathematician named Theodor Kaluza. In the 1930s, Kaluza and Klein said that space was not four-dimensional but five-dimensional, and canceling one of the five dimensions, there were four. His original thought was quite clear, and I can explain it to you. You know that physics has an electromagnetic field, which is written as a 4×4 matrix. This is one, two and three, which is the number of dimensional spaces. So, these three dimensions are the electric fields E_x, E_y, E_z, and the magnetic fields are H_z, H_y, H_x. Kaluza and Klein said that if one makes it bigger, into a 5×5 matrix. This is the fifth dimension. By moving it here, there are now no longer only four spatial dimensions, which, according to him, represent the vector potential. So now here came the vector potential. That is to say, he meant that the electromagnetic field comes from that fifth space—from five dimensions to four dimensions. That's why it is called a reduction of dimension.[7] What string theorists are now doing is reducing ten dimensions to four dimensions. As a Calabi–Yau space has six dimensions, so Yau Shing-Tung is involved.

Ji Lizhen: After I went to Northeastern University, I met Mark Goresky, Robert MacPherson, and then Armand Borel. Then I looked into Borel, and it seemed you were familiar with him.

Yang Chen-Ning: A colleague.

---

[7]See *A Note On Einstein, Bergmann, and the Fifth Dimension*. arXiv: 1401.8048. Chinese translation in *Science and Culture Review*, 5, 2017.

Borel and Compactification

Ji Lizhen: I mentioned him because after I graduated, I was interested in intersection cohomology, which was introduced by MacPherson and Goresky on singular space. There was the Zucker conjecture at that time, which asserted that, for the Hermite locally symmetric space, its Baily-Borel compactness is a singular space, and its intersection cohomology isomorphism is equal to its L^2 cohomology. It was a problem that Borel worked on for a while. Because of this, I became interested in Lie Groups. Then I went to Princeton and met Borel. I was Borel's last collaborator, and we wrote a book and an article together.

Yang Chen-Ning: But weren't you Borel's Ph.D. student?

Ji Lizhen: No.

Yang Chen-Ning: Borel was my colleague for many years.

Ji Lizhen: Seven years, I suppose—seven or eight years.

Yang Chen-Ning: He was there later than I. When I left there in 1966, he was still there. And you know, he then went to the University of Hong Kong for many years.

Ji Lizhen: Yes, I went there with him. I got to know him in 1997. Later, he told me that the University of Hong Kong organized activities, and he would invite me there. I also looked into it and found that Borel didn't have any students of his own.

Yang Chen-Ning: He had no students at the Institute for Advanced Study.

Ji Lizhen: So, I consider myself his student. Why? The reason is that after he died, I organized a big memorial event in his honor.

Yang Chen-Ning: So, you are equivalent to being his postdoctoral student. When did you go there (Institute for Advanced Study)?

Ji Lizhen: Between the years in 1994 and 1995. I never spoke to him. And once in 1996 or 1997, I went to Robert Langlands's 60th birthday. Before I went there, William Casselman told me that Borel used to run a seminar on compactification at the Institute for Advanced Study. Because I knew Casselman, and I told him I was doing compactification, I then went to see Borel. I knocked on the door, and he was inside. I went in and said, "I heard that you were interested in compactification." He said nothing. Then I talked to him for two hours. He became very interested, and said that he would organize an activity in Hong Kong SAR the following year. He asked me if I wanted to come.

Yang Chen-Ning: Which aspects of mathematics did you do with him?

Ji Lizhen: Well, the Euclidean plane, the hyperbolic plane and the sphere all have many symmetries. They are the most basic examples of symmetric spaces. Of the three symmetric spaces, the quotient space of the hyperbolic plane provides most of the symmetric spaces. The natural generalization of hyperbolic planes constitutes noncompact symmetric spaces, and they also have many quotient spaces. Among them, quotient spaces given by the action of arithmetic subgroups are especially important in the fields of number theory (automorphic form), algebraic geometry, differential geometry, and topology. For many applications, we need to compact the symmetric space and its noncompact quotient space. The book Borel and I published deals with these kinds of questions. For example, the hyperbolic plane can be transformed to be a unit circle or upper half plane in the complex plane. We can add its closure to the extended complex plane. For a symmetric space, there is no outer space, which is so obvious that we can get compactness directly by taking closures. I learned that the Borel–Serre compactification of locally symmetric spaces bears some resemblance to the Many-Body Problem. I remember you wrote a book on the

Many-Body Problem.[8] Why am I mentioning this? There's a mathematician at MIT right now named Richard Melrose. He is a leading figure in contemporary analysis. What he wanted to study was the Many-Body Problem, considering the asymptotic behavior of the Green function and, more generally, the asymptotic behavior of a resolvent kernel at infinity. His idea was to use some compactification to get some manifolds with sharp angles. He could understand the asymptotic behavior of these functions, so that he could study the Many-Body Problem based on geometry on the boundary. He then thought that his compactification shared properties similar to those of the Borel–Serre compactification of locally symmetric spaces from Lie groups and discrete groups. It is amazing and beautiful to see two completely different concepts so closely related to each other.

Yang Chen-Ning: What you are talking about is not the so-called dynamical systems theory developed by mathematicians, though.

Ji Lizhen: No. They are more interested in the asymptotic behavior of the Green function and, more generally, the asymptotic behavior of a resolvent kernel at infinity. This is related to the Many-Body Problem, right?

Yang Chen-Ning: The Many-Body Problem is of course most appealing to physicists, from two-body, three-body to multi-body. It has a close connection with the experiment, but it does not connect with mathematicians' so-called dynamical systems. Xia Zhihong works on this mathematician's dynamical systems, and he has come back to China.

Ji Lizhen: Xia Zhihong. I heard he went back to the Southern University of Science and Technology.

Yang Chen-Ning: I remember John Willard Milnor telling me 20 years ago that there was a young Chinese mathematician who had almost

---

[8]C. N. Yang. *The Many-Body Problem*. Lectures given at Latin American School of Physics, Centro Brasileiro de Pesquisas Fisicas, Rio de Janeiro, June 27–August 7, 1960.

solved the Painlevé conjecture. Later I realized it was Xia Zhihong. He studied at Northwestern University, and you at Northeastern University?

Ji Lizhen: Northwestern University is better.

Yang Chen-Ning: Do you know Wu Fayueh[9] from Northeastern University?

Ji Lizhen: I haven't heard of him. Does he come from the physics department?

Yang Chen-Ning: He is now the most important professor in the Department of Physics at Northeastern University and holds the title of University Professor. He has been there for many years and now is retired. This university now takes him very seriously because I think he is the most important physics professor at Northeastern University.

Ji Lizhen: I don't know about the Department of Physics, but in the Department of Mathematics, I know Terng Chuu-Lian, a postdoctoral student of Chern Shiing-Shen. The Department of Physics of Northeastern University is pretty good.

Yang–Baxter Equation

Yang Chen-Ning: The Department of Physics is generally not so strong, but he has done a good job. His theoretical physics is what the Chinese are good at. Why? Because they work so hard. The Chinese keep doing what no one else could do and make achievements. What he does might have a close relation with the Yang–Baxter equation.[10]

---

[9]Wu Fayueh (1932–2020) was a Chinese-born theoretical and mathematical physicist, as well as a mathematician, known for his research in solid-state physics and statistical mechanics.

[10]See Fa Yueh Wu, *Exactly Solved Models: A Journey in Statistical Mechanics*, World Scientific, 2009.

Yang Chen-Ning and Rodney Baxter. Taken on 22 May 1999, Stony Brook

Ji Lizhen: On the other hand, I find that symmetry actually is closely connected with your work.

Yang Chen-Ning: Yes.

Ji Lizhen: Because the Yang–Baxter equation is an integrable system, which has implicit symmetry.

Yang Chen-Ning: Of course. Because the Yang–Baxter equation, if you ask me, is really just a generalization of permutation groups. The difference is that the generalization of permutation groups doesn't have a function in it, whereas the Yang–Baxter equation has a function, which mixes up with the permutation group and becomes a cubic equation. There is more variety, and I can't figure it out now. You are not doing this kind of thing, are you?

Ji Lizhen: No, because I just heard that the Yang–Baxter equation might be relevant to integrable systems, and integrable systems have a lot of implicit symmetry, so I have to take a look. Why did I mention the Borel–Serre compactness? Because symmetry plays an important role in it. For Melrose, having some similar structure is equally important. Melrose is primarily interested in Laplace and Schrödinger operators. Since writing my dissertation, I have been interested in the Spectral Theory of the Laplace operator. Therefore, we have a good understanding of the Schrödinger equation. In short, the geometry behind the Many-Body Problem is very similar to the Borel–Serre compactness of locally symmetric spaces. And symmetry is important in all of them. I think this is surprising. You've written a book on the Many-Body Problem, and you've known Borel for a long time at the Institute for Advanced Study. That's why I thought I'd tell you.

Yang Chen-Ning: I think Lin Kailiang gave me a book on Marius Sophus Lie and Felix Klein,[11] and you wrote the preface. Did you edit this series of books?

Ji Lizhen: Yes, it was published by Higher Education Press (the Panorama of Mathematics Series, edited by academician Yan Jia'an and Professor Ji Lizhen).

Yang's Advice on Translation and Publishing

Yang Chen-Ning: Speaking of this, I have a piece of advice. I suggest that whenever you print a foreigner's name or place name, put it in italics. You can start with this series of books, and the good thing is that the word 费米, for example, looks like ordinary Chinese characters, not a person's name. Especially when I see a very long Russian name, I have to read it several times to figure out where it ends. If you put the name in italics, like katakana in Japanese, it immediately shows up, and you know it's a person's name. I don't think you need any government support for that, and you can do it yourself. Like the book

---

[11]Yaglom, *Felix Klein and Sophus Lie: Evolution of the Idea of Symmetry in the Nineteenth Century*, translated by Zhao Zhenjiang, Higher Education Press, 2016.

by Klein and Sophus Lie you just mentioned, there are a lot of names in it, especially those I haven't heard of, so I don't know if a name has three characters or four characters, and I have to read it twice. If a name is italicized, I can figure it out immediately. Here is another tip. If you have an index, don't sort by stroke numbers.

Wang Liping: We don't do that anymore. It is in alphabetical order now.

Yang Chen-Ning: Alphabetical order is much more convenient than stroke count.

Wang Liping: Yes, now it is in LaTeX and sorts in alphabetical order. In the 1950s and 1960s, we used to use the number of strokes, and we couldn't count them out.

Yang Chen-Ning: Yes, and when you use strokes, foreigners' names and Chinese names are all mixed up together. And if you use alphabetical order, it's natural to include both Chinese and English names without any problems.

Wang Liping: Yes, your suggestions are good, especially the first one. We haven't noticed this problem before.

Ji Lizhen: The suggestions are all good. You're familiar with the work of Klein and Sophus Lie, aren't you? The theories they developed have great applications in physics.

Klein, Sophus Lie and Their Conflict

Yang Chen-Ning: I actually flipped through this book, and there are many things in it that I don't know. I thought, for example, that group theory was developed by Évariste Galois. In fact, it is now generally accepted that group theory began with Galois, but it was the book by Camille Jordan that made this understood. So, I didn't know about this before. In addition, Jordan taught Sophus Lie and Klein. The influence of Jordan's book on mathematics was enormous. Another thing I had not known was that Sophus Lie was not very pleased with Klein.

Ji Lizhen: So, I wrote in more detail, which many people had previously not known.

Yang Chen-Ning: I actually think that Klein might be a difficult person to deal with, though the book doesn't seem to say much about Sophus Lie's personality.

Ji Lizhen: Yes, I wrote another article about the life and work of Sophus Lie called "Sophus Lie: A Real Giant in Mathematics," translated by Lin Kailiang. Lin Kailiang can send it to you then.

Yang Chen-Ning: Where has it been published?

Ji Lizhen: Yes, it is in English. I'll show you the Chinese version.

Yang Chen-Ning: There was something else I didn't know until I read this book. I thought that Sophus Lie specialized in the study of Lie transformation groups. This is not wrong, but I didn't know that he had already come up with the concept of Lie algebra. In other words, it is Lie himself who developed this Lie algebra, rather than other people who supposedly developed Lie algebra from Lie groups.[12]

Ji Lizhen: Yes, it was a very important step for him because it (Lie algebra) simplified these (Lie groups). Lie algebra is one of his most important contributions.

Yang Chen-Ning: He was not able to classify this Lie algebra like Elie Cartan, and this was his failure.

Ji Lizhen: This was done by Wilhelm Killing.

Yang Chen-Ning: It was not done until Killing.

---

[12]In fact, Killing also independently proposed Lie algebra.

Ji Lizhen: Yes, and because of this, Sophus Lie and Killing got on very badly.

Wu Fan: Yes, very badly. They joked that "Lie is going to kill Killing."

Yang Chen-Ning (laughs): Was Killing a student of Sophus Lie?

Ji Lizhen: No.

Wu Fan: Killing was a student of Weierstrass.

Sophus Lie and Felix Klein: The Erlangen Program and Its Impact in Mathematics and Physics

Yang Chen-Ning: A student of Weierstrass? Is he German? Why didn't Killing create what Elie Cartan did?

Ji Lizhen: Elie Cartan clarified what Killing had done (the classification of Lie algebra).

Yang Chen-Ning: Was Elie Cartan later than Killing? Why didn't Killing do what Elie Cartan did?

Ji Lizhen: Because then Killing stopped doing mathematics and abandoned it. Killing was doing this when he was a high school teacher; however, he stopped doing mathematics when he got a professorship. You see, here is the book I edited.

Yang Chen-Ning: Is the book available?

Ji Lizhen: Yes, it's published by the European Mathematical Society. I'll send you one when I get back.

Yang Chen-Ning: You can email me the name of the book.

Ji Lizhen: OK, it includes two of my articles. One is a preface about Klein, and the other is about the life of Sophus Lie. Because I'm interested in Lie groups but didn't know anything about the life of Lie or Klein and couldn't find the right one, so I wrote it myself. I did another inquisitive thing the year before. I thought that since Klein's influence was so great, I would walk around all the places where Klein had worked. I went first to Erlangen University where he had taught, next to Munich where Klein had stayed, and then to Göttingen and Leipzig. I thought it was very interesting. Klein is amazing.

Yang Chen-Ning: Well, well … Did Klein get into a fight with Poincaré?

Klein and Poincaré's Famous Dispute on the Uniformization of Compact Riemann Surfaces

Ji Lizhen: Yes, there was a dispute. Then, Klein and Poincaré quarreled with each other. As a result, Klein's later research was scrapped.

Yang Chen-Ning: You said that happened after the dispute. Did it prove that Poincaré was better than Klein?

Ji Lizhen: Because Klein was exhausted. At his peak, this young man suddenly appeared. Klein, who was the best mathematician in Germany at that time, didn't want to give up and was determined to struggle to compete with Poincaré. Right after he published his final work, he was on the verge of collapse.

Yang Chen-Ning: I once read one of his articles, which stated that German mathematics and French mathematics were different.

Ji Lizhen: Yes, it was published by Higher Education Press and included in *The Development of Mathematics in the Nineteenth Century*.[13]

Yang Chen-Ning: The tone of his writing was that Germany was superior to France in mathematics. I've read this article.

Ji Lizhen: It's very interesting. Klein wrote dozens of letters when he quarreled with Poincaré, and now I'm asking Lin Kailiang to have these translated into Chinese.[14]

Yang Chen-Ning: Do you mean that you have found the letters?

Ji Lizhen: They are in Poincaré's collected works, translated by Poincaré's grandson (François Poincaré).

---

[13]Felix Klein, *The Development of Mathematics in the Nineteenth Century*, translated into Chinese by Qi Minyou, Higher Education Press, 2010.

[14]The English version was included as an appendix in *Uniformization of Riemann Surfaces*, ENS Editions, Lyon, 2010, edited by Henri Paul de Saint-Gervais.

Yang Chen-Ning: But were these letters written in German?

Ji Lizhen: That's right. Klein wrote them in German, and Poincaré's grandson translated them into French.

Yang Chen-Ning, Courant and Courant's Son

Yang Chen-Ning: Well, they're all much older than me, but there's a guy in Göttingen I know named Richard Courant. He's about 30 years older than me. But I know him very well, and it's because I know his son very well. His son is a very important theoretical physicist named Ernest Courant, who is two years older than me and still alive. He contributed so much that I nominated him several times for the Nobel Prize because he invented the strong focusing principle. What is the strong focusing principle? You know, in high-energy physics you have to make big accelerators, and these are becoming bigger and bigger. The one (LHL) at CERN (European Organization for Nuclear Research), is a curved underground ring, whose radius is more than 20 kilometers. The accelerator first began to be built right after the war. The United States had money then, so they built the biggest accelerator.

In order to have an accelerator generate high energy, its radius must be large. Because particles grow unstable during the process, they oscillate on top of the accelerator. A circle, let's say, of this space, when it is 100 meters in diameter, has oscillations. However, when it goes from 100 meters to 10,000 meters, it oscillates proportionally a lot more, so the vacuum tube has to get wider. By the 1950s, the largest particle accelerator in the United States was called Bevatron, and its diameter, I think, was three or four kilometers. Its vacuum tubes were so wide that you could climb into them. The larger the radius, the larger the vacuum tubes had to be. In other words, the cost is a cubic function of the radius, because the bigger the front side, the bigger the cross section. It becomes a cubic relationship, so you can't make an accelerator very big.

In 1952, Ernest Courant and two other people wrote a paper[15] and

---

[15]Courant, E. D., Livingston, M. S., Snyder, H. S. (1952), "The Strong-Focusing Synchrotron: A New High Energy Accelerator," *Physical Review*, 88 (5): 1190–1196.

invented something called the strong focus principle. This principle tells that if you compress the oscillation in a certain way, and the energy gets bigger, your cost becomes linear depending on the radius, not cubic depending on the radius. So today you go to CERN, and it's three or four kilometers in diameter, yet the vacuum tube is very thin. It's because of their work that this can be done. When I nominated him for the Nobel Prize, I said that without them, you wouldn't be able to do this right now, and high-energy physics would come to a halt. That research is very important. This actually has a close relationship with Jürgen Moser's dynamical system theory and Vladimir I. Arnold's KAM theory.[16]

You could also say that when they put forward this concept, those people who were doing KAM theory were very happy, so Ernest Courant became very important. I knew him first, and then his father. After I met his father, I also met Kurt Friedrichs and the woman who wrote the biography of David Hilbert. Isn't she Constance Reid?[17]

Ji Lizhen: Constance Reid. Do you also know her?

Constance Reid and Hilbert's Biography

Yang Chen-Ning: Do you know why Constance Reid wrote a biography of Hilbert?

Ji Lizhen: No, I don't know. Is it because she's Julia Robinson's sister?

Yang Chen-Ning: With the help of her sister, a famous mathematician, she wrote a biography of Hilbert. In writing his biography, she visited Courant and Friedrichs. Once it became a success, she wanted to write a biography of Courant.

Ji Lizhen: I read all her biographies. Because of her biography of Hilbert and Courant, I thought I must go to Göttingen.

---

[16]KAM theory is named after A. N. Kolmogorov, V. I. Arnold and Jürgen Moser.
[17]Constance Reid is best known for her four biographies of mathematicians: *Hilbert*, *Courant in Göttingen and New York*, *Neyman*, and *Julia: A Life in Mathematics*. The first three have Chinese translations.

Yang Chen-Ning: What's your accent?

Ji Lizhen: Wenzhou.

Yang Chen-Ning: Are you from Wenzhou, then?

Ji Lizhen: Yeah, from Wenzhou. I do have a strong accent, don't I?

Yang Chen-Ning: Yes, you sound a little like Gu Chaohao.[18]

Ji Lizhen: Yes, he's from my hometown. He always says I'm not good at Chinese. I don't think he is either.

Yang Chen-Ning: There aren't many people surnamed Ji, are there?

Ji Lizhen: No. But our surname has a relatively early origin. I went to Wenzhou a few days ago. In our ancestral temple, the memory of our ancestors can also be traced back thousands of years ago, though I cannot remember the time clearly.

Yang Chen-Ning: Did you grow up in Wenzhou?

Ji Lizhen: Yes, I grew up in Wenzhou.

Yang Chen-Ning: Because you still have a strong accent.

Ji Lizhen: Really? Yes, I went to middle school in Wenzhou, and then I went to Hangzhou University.

Yang Chen-Ning: Were you admitted to Hangzhou University?

Riemann's Manuscript

Ji Lizhen: Yes, I studied at Hangzhou University. It's a coincidence

---

[18]Gu Chao-Hao (1926–2012), a mathematician from Wenzhou, Zhejiang Province, was one of the collaborators of Mr. Yang Chen-Ning.

because I didn't use to like mathematics, and in fact, mathematics was my worst subject among mathematics, physics and chemistry. Anyway, for some inexplicable reasons, I went there studying mathematics. I don't speak either Chinese or English very well, and I've forgotten how to write Chinese. In fact, I can't write Chinese. It was because I had read two of Constance Reid's books that I thought I must go to Göttingen, and I went there twice. Last time I went to Göttingen, I did something quite special. I said I had to see Bernhard Riemann's manuscript, because the historical library in Göttingen had Gauss's and Riemann's manuscripts. I was very surprised when I saw Riemann's manuscript, because we used to say that Riemann's papers had very few equations and calculations, and a lot of people said that Riemann was a person who thought in terms of concepts. It was as if he could just sit there and think, which was so amazing. All the pieces of Riemann's paper were collected, page by page.

Yang Chen-Ning: Is it still in Göttingen?

Ji Lizhen: Yes, in Göttingen. Riemann's manuscript *On the Fundamental Assumption of Geometry* is still there.

Yang Chen-Ning: Handwritten?

Ji Lizhen: Yes. Anyway, I was very excited to go to Göttingen, and I even found the houses where Courant, Klein and Hilbert lived.

Hermann Weyl's House in Princeton

Yang Chen-Ning: Do you know Hermann Weyl? In 1933, he built a house on Mercer Street in Princeton. Then he died in 1955, and according to the rules of the Institute for Advanced Study at that time, because his house was built on the land of the Institute for Advanced Study, when you sold it, you had to sell it to the institute first. So, it was sold to the institute. The house was sold back in about 1956, and I bought it from the Institute for Advanced Study in 1957, so I

Weyl—Yang Chen-Ning's former residence

lived there.[19] The house was at 284 Mercer Street, within walking distance of the institute. I lived there for nine years, from 1957 to 1966, and then moved to Stony Brook, New York. When I left, I sold it back to the institute. It happened to be that at that time, in 1966, Julius Robert Oppenheimer was retired and wasn't the director of the institute anymore. His cancer was in the third stage, so he moved from the director's house to this house. After I left, Oppenheimer moved in a couple of months later. He lived there only for half a year before he died. The house was then sold back to the Institute for Advanced Study, and it was occupied by a physicist named Marshall Rosenbluth for a while. Rosenbluth was my schoolmate at the University of Chicago, and he was, I think, the most important physicist in plasma physics in the past few decades. When he left, he sold the house to the Institute for Advanced Study. And every time I went back to Princeton,

---

[19]Dyson wrote in his famous Einstein public speech *Birds and Frogs*: "Soon after Weyl left Princeton, Frank Yang arrived from Chicago and moved into Weyl's old house. Yang took Weyl's place as the leading bird among my generation of physicists."

I went there. The house is still there, but the color has changed. It was yellow when we were there. Weyl had it built in the Art Deco style. Do you know Art Deco? It was a very popular style of architecture in the 1930s. This house is square, and the windows are square. There's something special about it, and I'll send you a picture.

Ji Lizhen: Did Borel also live on Mercer Street?

A Researcher's House at the Institute for Advanced Study at Princeton

Yang Chen-Ning: Because all around us was the land of the Institute for Advanced Study, Borel lived next door to me, and Dyson was my close neighbor. Borel is dead now. I think the house has been sold.

Ji Lizhen: No, his wife still lives there.

Yang Chen-Ning: Does she? His wife, I think, is 90 years old.

Ji Lizhen: She was 80 when Borel died in 2003, and now she's in her 90s. It seems that his wife was a little older than him, so she must be 95 or 96 now.

Yang Chen-Ning: He's of the same generation as me.

Ji Lizhen: Yes, because my kids used to have contact with Borel.

Borel: A Serious Scholar

Yang Chen-Ning: I can also tell you that Borel (I don't know how he usually got along with mathematicians) didn't talk much with our physicists because he, you could say, pursued accuracy too much.

Ji Lizhen: Yes, very precise.

Yang Chen-Ning: Then one day in the 1960s, I came to him and said, "Now it is said that everyone is studying topology. Could you explain to us physicists what topology is?" He declined, and said he couldn't explain it. Later on, several of us desperately demanded an explanation, and he agreed. So, for about a week, every other night, in a room with Borel alone, three or four physics professors from the Institute for Advanced Study listened to him. We found that he tended to do everything to very high dimensions. We said that we didn't understand these high-dimensional cases, and we hadn't learned this. We asked him to make it into lower dimensions for us. He declined and said that when it was in lower dimensions, it lost all its charm. After a long argument, he finally got down to the lowest dimension. I remember clearly that the first thing he explained to us was Brouwer's fixed-point theorem. I'd never heard of it, and it turned out I could prove the fixed-point theorem, which is circles … This was wonderful, and the 'hairs' represent similar cases. (Dr. Yang asked while touching his head).[20]

Ji Lizhen: Yes, the vector field on a sphere.

Yang Chen-Ning: So, this was the first time I'd heard about this kind of topology.

Ji Lizhen: It's beautiful. Vector fields have to do with algebra …

Morse Theory and Physics

Yang Chen-Ning: In the article I just gave you, I talked about the first application of topology in physics, a story that physicists and mathematicians do not know.

Ji Lizhen: Really? What is it?

Yang Chen-Ning: Morse theory.

---

[20]Hairy ball theorem. It asserts that there is no continuous unit vector field in a two-dimensional sphere.

Ji Lizhen: Morse theory? Do you mean that Morse theory comes from physics? This was the first formal application of topology to theoretical physics, Morse theory. What dimension of space does Morse theory use?

Yang Chen-Ning: Two and three dimensions. Look at the article. It's actually very simple. There was a postdoctoral student of my time, Léon van Hove. Later, he became the director of CERN. At that time, many people used computers to calculate, and they found that some curves had some singularities. People did not understand at that time, so they thought that his calculations were wrong. Then van Hove said, not only did he get it right, but there had to be a singularity. This was Morse theory, and that was the first time I had heard about it.

Ji Lizhen: Did Morse also worked at the Institute for Advanced Study?

Yang Chen-Ning: Morse was a professor at the Institute for Advanced Study, but I didn't know that then. I only knew that Morse theory was very important.

Ji Lizhen: People in the mathematics department didn't have much contact with those in the natural science department, did they? That's interesting.

Yang Chen-Ning: Morse was from an older generation.

Ji Lizhen: Yeah, he's older. So, you have more contact with Weyl.

Weyl's Disfavor of Hua Loo-Keng

Yang Chen-Ning: I ended up living in his house. Although Weyl and I didn't talk about mathematics and physics at the Institute for Advanced Study. I used to talk a lot with him at parties. Every time he saw my wife and me, he would speak ill of Hua Loo-Keng.

Ji Lizhen: Really? Why?

Yang Chen-Ning: He doesn't like Hua Loo-Keng.

Ji Lizhen: Oh, really?

Yang Chen-Ning: I don't know why. He just doesn't like Hua Loo-Keng.

Ji Lizhen: Oh, I see. Is it probably because Hua Loo-Keng did analytical number theory? Weyl may …

Yang Chen-Ning: Chern Shiing-Shen and Hua Loo-Keng were both invited to the institute by Weyl. Later, he did not like Hua Loo-Keng for some unknown reason. Lin Kailiang, according to the information you showed me last time, when Weyl wrote to invite them to the institute,[21] he first put forward that he thought they were the two best mathematicians from China. In the letters, he pointed out that they were different in their behavior and in doing mathematics. So, I think in the early 1940s, he (foresaw) … These letters are very interesting.[22]

Ji Lizhen: I'll take a look. Weyl is very interesting! I read his article on Emmy Noether, and remember what he said about Noether.

Yang Chen-Ning: I'll send you an email later. You may ask Lin Kailiang about this. I wrote an article about the 100th anniversary of Weyl's birth in 1985, and there was a meeting in Zurich to honor him.

---

[21]On March 24, 1943, Weyl wrote to Professors Alexander, Einstein, Morse and Oswald Veblen at the Institute for Advanced Study at Princeton, recommending Chern Shiing-Shen and Hua Loo-Keng come to visit: "To the best of my knowledge, the two most outstanding Chinese mathematicians are Chern Shiing-Shen and Hua Loo-Keng (of National Tsing Hua University in Kunming). The latter contributed significantly to the Hardy–Littlewood–Vinogradov school of analytic number theory. In a manuscript he recently sent me, he repeated much of the work of Carl Ludwig Siegel in his masterpiece on symplectic geometry. It is important for Hua Loo-Keng to have close contact with Siege. Whether or not we could invite him to the Institute for Advanced Study at Princeton, I think the second-best candidate from China was Hua Loo-Keng, not Xiong Quanzhi."

[22]See J. W. Richard and Yuan Hong, Hua Loo-Keng and Weyl, included in "Mathematics and Humanities," series no. 12, *One Hundred Years of Mathematics*, edited by Shing-Tung Yau *et al.*, Higher Education Press, 2014.

Ji Lizhen: I've read this book,[23] and Borel's article is also in it.[24]

Yang Chen-Ning: Yes, there is Borel's article, and there is my article. Have you read my article?[25] Weyl, of course, has a great influence on physics.

Ji Lizhen: I think, thanks to Weyl's work, group theory became important later. This is because Sophus Lie's previous work was relatively fragmented, while Weyl and Elie Cartan integrated Lie's work.

Group Theory of Weyl, Cartan and Lie Groups

Yang Chen-Ning: The method of calculating the characteristic indices of these typical groups is absolutely fantastic.

Ji Lizhen: Have you read his book *The Classical Groups*?[26]

Yang Chen-Ning: I haven't read his book carefully, but the methods he used came out later.

Ji Lizhen: Really? Yeah, it seems to be a very famous book. I haven't read it. I should make time to read it.

Yang Chen-Ning's Article on Weyl and Symmetry

Yang Chen-Ning: If you read my article about Weyl, you would know that he had a very odd approach, which I mentioned in that article. Because in 1929, he wrote an article about three kinds of symmetry in physics. One was P (parity), one was C (charge conjugation) and

---

[23]Hermann Weyl, 1885–1985: *Centenary Lectures*, Springer-Verlag, 1986.

[24]Borel's article was titled "Hermann Weyl and Lie Groups," and included in the aforementioned book.

[25]"Weyl's Contribution to Physics," included in *Dawn Collection* by Yang Chen-Ning and Weng Fan, SDX Joint Publishing Company, 2008.

[26]*The Classical Groups: Their Invariants and Representations*, Princeton University Press, 1939.

one was T (time reversal). He put these three symmetries together and wrote a paragraph about it,[27] which was a very rare thing because at that time, no physicist would do this. But what's strange is that, after more than ten years, it turns out that these three are very important, and they all have names called P, C and T. So, when Wu Chien-Shiung's experiment came out, the importance of CPT came out. In other words, by 1957, it was suddenly clear to everyone that there was something very deep going on here, something that Weyl had already discussed in 1929.

Ji Lizhen: So, does this have something to do with the parity non-conservation paper for that you and Lee Tsung-Dao won the Nobel Prize?

Yang Chen-Ning: Of course. Mirror symmetry is Weyl's idea (editor's note: Eugene Paul Wigner of the same period independently proposed the idea in 1928).

Ji Lizhen: Weyl mentioned this idea in the 1920s.

Weyl, Abelian Gauge Field Theory and Non-Abelian Gauge Field Theory

Yang Chen-Ning: So, Weyl was both a mathematician and a philosopher. His philosophy was so accurate that he started the normative field. All I did was generalizing his gauge field from commutative to non-commutative. This generalization was extraordinary. After seven years of studying, it turned out to be very correct and very simple.

Ji Lizhen: Your gauge field is very important, so elementary particles all depend on ...

---

[27]In the preface to his 1930 edition of *The Theory of Groups and Quantum Mechanics*, Weyl said: "The fundamental problem of the proton and the electron has been discussed in its relation to the symmetry properties of the quantum laws with respect to the interchange of right and left, past and future, and positive and negative electricity. At present no solution to the problem seems in sight; I fear that the clouds hanging over this part of the subject will roll together to form a new crisis in quantum physics."

Yang Chen-Ning: Now it's sweeping the field. Veltman, for example, won the Nobel Prize because he worked on gauge theory. He and his student Gerardus't Hooft proved that noncommutative gauge theory could be renormalized, and that's why they won the Nobel Prize in Physics (in 1999).

Ji Lizhen: Because I knew that symmetry was important in your gauge field theory with Mills, I suddenly realized that it was also important in the Yang–Baxter equation yesterday, so symmetry was very important in your work.

Yang Chen-Ning's First Contact with the Concept of Symmetry

Yang Chen-Ning: Yes. The first thing I knew about symmetry was that when I was in middle school, there was a magazine called *Middle School Students*, published by Kaiming Bookstore. I think you can still find it in the library. One of the authors of *Middle School Students* is Liu Xunyu.[28]

One day—I was probably in grade two or grade one in junior high school—I read his article explaining the distinction between odd permutation and even permutation. It was wonderful, and I first came to know about the little wonderful things of symmetry from that article.[29]

---

[28]Liu Xunyu's popular science books also influenced Hsu Leetsch C. and Gu Chaohao. In the article "Talk About My Experience and Feelings about Learning Mathematics in My Youth," Hsu Leetsch C. said, "Liu Xunyu's *Mathematical Interest* aroused my interest in mathematics. This book was published by Kaiming Bookstore which was established before the founding of the People's Republic of China. When I was in junior high, I read this book and I saw what mathematical induction was all about." Gu Chaohao looked back to 1937 when he studied in Wenzhou Middle School and said, "I remember when I was reading Liu Xunyu's *The Garden of Mathematics*, there was a paragraph about calculus thought, starting from 'what is speed'. I thought I knew a lot about speed, acceleration, etc., but after I read this book, I found that the real concept of speed can only be understood with calculus, so I became more and more interested in mathematics."

[29]In his article "Some Suggestions on How to Learn Science" (included in *Yang Chen-Ning Anthology*), Yang Chen-Ning said:

As early as middle school, I became interested in mathematics by chance

Ji Lizhen: Really? There are some popular and elaborative articles that are very important and enlightening to young people.

Yang Chen-Ning: He's very good at writing. There was one article by him every few issues. I don't know whether he majored in mathematics.

Lin Kailiang: He did. I've read about him.

Yang Chen-Ning: Did he write college textbooks or something like that?

Lin Kailiang: Maybe. I need to check this out. It seems that he graduated from Beijing Normal University.[30]

Ji Lizhen: Speaking of symmetry, last time I looked at Sophus Lie's

---

and discovered my mathematical ability. In the 1930s, Kaiming Bookstore published a magazine called *Middle School Students*. I think some libraries in Hong Kong SAR still have this magazine. It's a very good magazine aimed at middle school students. It's serious and interesting. There is Liu Xunyu, a mathematician who has written many easy-to-understand and extremely interesting articles on mathematics. I remember I read an article of his on an intelligence quiz and learned about the most important mathematical concepts of permutation and odd-even permutation. You may have seen this intelligence quiz. It divides a box into 16 squares, 15 of which are filled with a cube and one is empty. You can move the cubes back and forth. The cubes start out haphazard, requiring you to move them into a particular state. This is a very interesting conundrum. I'm sure you've seen it, and it must be very popular in Hong Kong SAR. However, how to move and whether to get to the desired state from the initial state requires mathematical analysis using the concept of even-odd permutation. If someone has a mathematical talent and enjoys solving this kind of problem, they are lucky. He or she should cultivate this interest by delving as far and wide as possible.

This game is called 15-Puzzle and is introduced in the second chapter of Chen Jingrun's *Introduction to Combinatorial Mathematics*.

[30]Liu Xunyu was the classmate of Yang Chen-Ning's father, Yang Wuzhi, at Beijing Normal University and later went to teach at Guiyang Middle School. During the Southwest Associated University period, he taught at Southwest Associated University's Normal College. He has written many excellent popular science books and has great fame. The most influential books are *Mr. Ma Talks Arithmetic*, *Mathematical Interest* and *The Garden of Mathematics*. His readers include Hsu Leetsch C., Gu Chaohao, Qi Minyou and Feng Zikai (Wu Wenjun is also said to be among his readers). He has also written a textbook on calculus.

biography, he always felt that when he was alive, no one in the world recognized him. Then he thought, "I firmly believe that what I do will be recognized."

Langlands Program and Symmetry

Yang Chen-Ning: Well … Is there a program now called the Langlands program? Do you understand what the Langlands program is? Does it have anything to do with symmetry?

Ji Lizhen: Yes, it's very important.

Yang Chen-Ning: What is the relationship?

Ji Lizhen: A very important problem in number theory is solving equations, which is a field extension, equivalent to a Galois group. In the past, Leopold Kronecker and Heinrich Weber gave a very good explanation that for every given field extension, there is a group, and this is because equations have groups, and field extensions also have groups. Kronecker and Weber's theorem says that if the group is Abelian, there's a good structure theory. So now what everybody wants to know is, what to do if this group is non-Abelian, and if this Galois group is non-Abelian, how do we describe this field extension?

Yang Chen-Ning: Why is this related to compactification?

Ji Lizhen: Well, how to know about this Galois extension … Given this, they have a function called L, because it describes the number of primes …

Yang Chen-Ning: Do you mean that L function is a generalization of *zeta* function?

Ji Lizhen: Yeah. So, for every given field extension, there's a Galois group, and for every formulation, there's an L function. Langlands said that, analytically, for locally symmetric spaces, it is also possible to construct an L function in modular form, and he concluded that the two L functions are equal, hence there is this program (Langlands Program[31]). Because of this program, the difficult things in number theory can be done by analytics, and the difficult things in analytics can be done by number theory. Why did Langlands study this? He seemed to say that …

Oh, yes, we published a collection of essays for Langlands, and he wrote us a reminiscence of thirty pages. He said when he just started to work at Princeton, Salomon Bochner encouraged him to teach a course in number theory which was called class field theory. Langlands had never learned it, but he stuck to it and got on with it.

Langlands' Lecture Standard

Yang Chen-Ning: But Langlands didn't know how to deliver a lecture.

Ji Lizhen: His lectures were difficult to understand.

Yang Chen-Ning: Yes, because I served on a selection committee of the Shaw Prize for many years, and one year he won the Shaw Prize.

Ji Lizhen: How did he speak?

Yang Chen-Ning: All the Shaw Prize winners have to give a lecture for a layman audience for one hour. He was given an hour, and he was the worst among all the 11 Shaw Prize winners for mathematics. You had no idea what he was talking about, and he had the habit of writing a few words in a corner and erasing them while he was still blocking the words so that nobody could see them.

Ji Lizhen: Bad habits.

Yang Chen-Ning: When you asked him a question, he didn't know what your question meant.

---

[31]A generalized introduction to the Langlands program can be found in Edward Frenkel's popular science book *Love and Math*, 2014.

Ji Lizhen: Yes, his articles have long sentences and are difficult to read.

Yang Chen-Ning: I heard that the Langlands program first appeared in a letter he wrote to André Weil.

Ji Lizhen: Yes, it's about to be translated into Chinese.

Yang Chen-Ning: Translated into Chinese?

Ji Lizhen: Yeah. It's written in German.

Yang Chen-Ning: In German? Did you say he wrote to André Weil in German?

Ji Lizhen: Yes, it's in German in the last part. I'll check if it's in English in the first part. Maybe it's in German.

Yang Chen-Ning: So, is it possible that he was influenced by Riemann?

Ji Lizhen: He's a bit of a weirdo, and now he's learning Chinese very slowly. I think he's written in Turkish because he likes to try different things. Letters from Langlands to André Weil include some articles about how he put forward the Langlands program, explanations of its application, and his life. They are collected in a book and translated into Chinese. The translation is almost ready, and it will probably be published early next year.[32]

André Weil is a very interesting man.

The First Meeting of Yang Chen-Ning and André Weil

Yang Chen-Ning: I know André Weil very well. Why? Because he was a professor at the University of Chicago, and then around 1960, he and Chern Shiing-Shen left. One went east, and the other went west, and he said to Mr. Chern: "I know why I went east (editor's note: the

---

[32]*Langlands Program and His Mathematical World*, Ji Lizhen collection, translated into Chinese by Li Jinghui *et al.*, Higher Education Press, 2018.

Institute for Advanced Study at Princeton), and you went west (UC Berkeley). The reason is that it's closer to France in the east and closer to China in the west."

Ji Lizhen: Ha, ha, that's a good explanation!

Yang Chen-Ning: Then he heard from Mr. Chern that there was a man named Yang Chen-Ning, so he came to the Institute for Advanced Study to find me. I was surprised when he came to me. He talked to me for about an hour in my office, and I found that his main intent was that there was a professor in the history department of the Institute for Advanced Study named Andrew E. Z. Alföldi. He said this guy was not good and asked me to gang up on him and kick him out.

Ji Lizhen: Really? Why?

André Weil's High Standards and Strong Personality

Yang Chen-Ning: I was shocked. We and those in the history department didn't even know what Alföldi did. I don't think he would have done badly if he was at the Institute for Advanced Study. He's fine, so why was him kicked out? I thought, "I'm not doing this with him. I don't know what André Weil did." But he liked to meddle, and as a result, he had a conflict with Oppenheimer, a big conflict. After the conflict with Oppenheimer, all of us in the Physics Department helped Oppenheimer, so André Weil didn't like me.

Ji Lizhen: Really? It's interesting. I have just read a book by his daughter Sylvie Weil.[33] It's not easy being his daughter, she said.

Yang Chen-Ning: He's very strange.

Wu Fan: It's in English. The letters from Langlands to André Weil are in English. The subsequent letters are written in German.

---

[33]Sylvie Weil, *At Home with André and Simone Weil*, translated by Benjamin Ivry, Northwestern University Press, 2010.

Yang Chen-Ning: To whom did he write?

Wu Fan: Letters to André Weil, written in English.

Yang Chen-Ning: Did his daughter write the letters for him?

Wu Fan: No, Langlands wrote to André Weil to explain Langlands's conjecture.

Ji Lizhen: Yes, then he wrote another one in the 1980s in German.

Yang Chen-Ning: Did you say it was written by Langlands, so it's in English? Oh, please (Wu Fan) send this to me by email.

André Weil's Sister and Their House in Paris

Ji Lizhen: I was wrong. He wrote one later in German. I was in Paris not long ago, and I went to the house where André Weil grew up. There was a brass plaque hanging on the outside of the house, and there was information about André Weil's sister, Simone Weil, but there was no word about André Weil.

Yang Chen-Ning: His sister is more famous than him.

Ji Lizhen: It seemed that he was upset and vented to his parents about it.

The Friendship Between Chern Shiing-Shen and André Weil?

Yang Chen-Ning: Dr. Chern told me that he was André Weil's best friend among mathematicians.[34] Then Dr. Chern said that in his later years, at first, they telephoned once every one or two months. It didn't

---

[34]When Springer published the collected works of Chern Shiing-Shen, André Weil wrote a wonderful preface, which was later translated into Chinese by Yang Chen-Ning himself. See "My Friend—Geometrist Chern Shiing-Shen," *Nature*, no. 8, 1979.

work later because they both had hearing problems. They couldn't hear each other via telephones, so they wrote letters later. I'm afraid that Dr. Chern was one of his few friends in mathematics.

Ji Lizhen: Yeah, he seemed to be very harsh to others.

Yang Chen-Ning: They knew each other when Elie Cartan was still alive. It must have been around that time, Dr. Chern was very Chinese, so it was easy for him to get along with people. This was clearly written in Weil's letter. Therefore, André Weil and Dr. Chern got to know each other very quickly. Dr. Chern went to Princeton in 1943, when André Weil was at Lehigh University. Why was he at Lehigh?

Ji Lizhen: He couldn't find a job.

The Process by Which Chern Shiing-Shen Obtained the Intrinsic Proof of the Gauss–Bonnet Theorem

Yang Chen-Ning: At that time, he had not been accepted by the Institute for Advanced Study and American universities, so he was very unhappy with the United States for all his life. Even though the United States invited him to come and work as a professor later, he was still not satisfied. This is why, when he was at Lehigh, Dr. Chern sometimes went to see him, and he showed Dr. Chern his article. I suppose the reason why Dr. Chern could write less than six pages only in a few weeks was because the proof of André Weil's paper was very complicated. But he had included an integrand, which, in physicists' terms, is the dot product of F and $\tilde{F}$. Dr. Chern had been trying to extend the Gauss–Bonnet formula to higher dimensions, but at the time he did not know the integrand of four or six dimensions. As soon as he saw André Weil's paper, he ignored how Weil had proved it, used the exterior differential form himself, and proceeded to write the paper made up of fewer than six pages in total, which, as the English expression captures well, "and created history."

In fact, it's a very interesting story. You could say the reason why Dr. Chern did this was that he had a very early appreciation of the beauty of external differential form and the Gauss–Bonnet theorem. He had already put these two together in two dimensions, so when he was at the Southwest Associated University in Kunming, he could prove the Gauss–Bonnet formula in two dimensions very simply. At that time, the problem of four dimensions was a well-known one. Everyone was engaged in tackling it, including Dr. Chern. But he didn't know what the integrand was, so when he saw it in the article by André Weil (and Carl Barnett Allendoerfer), he successfully worked it out immediately.

Ji Lizhen: That's lucky, too.

Yang Chen-Ning: The lesson is: you have to be interested in unsolved but solvable problems first, and with a suitable opportunity you can solve them.[35]

Ji Lizhen: Yeah, it's like plowing a field; the soil is ready, and the seed just falls in.

Yang Chen-Ning: Yes.

Ji Lizhen: Very interesting! It's funny that there is so much we don't know! I always thought that regarding certain mathematicians and physicists, we will be more interested in their theoretical results if we know more about the actual people and the stories behind them because their theoretical results then become much more enriched and more vivid.

Yang Chen-Ning: I've been in the United States for many years, so I know a lot of people who are doing mathematics, like Deane Montgomery.

Ji Lizhen: I've heard of the name, Montgomery, but I've never seen him or his work before.

---

[35]Interest→preparation work→breakthrough point are the three steps of research repeatedly emphasized by Yang Chen-Ning in his article "My Learning and Research Experience."

Montgomery and André Weil's Princeton War

Yang Chen-Ning: Montgomery became a professor at the Institute for Advanced Study early on because he solved the fifth problem of Hilbert. But as a person, well … Back then, both he and André Weil ganged up together opposing Oppenheimer strongly.

Ji Lizhen: What was the reason for their objections? Was it because of politics?

Yang Chen-Ning: They had a lot of reasons. I think Montgomery felt that Oppenheimer was not solely devoted to academic work. Does that make sense? I think it makes sense. André Weil and Montgomery, especially Montgomery, were interested in mathematics and nothing else; André Weil was interested in many things, but he didn't pursue them and focused only on his own mathematics. Oppenheimer was a scholar, and later became the director of the atomic bomb program. This led to him becoming one of the most important men in the United States, and he often went to Washington. Because of this, André Weil and Montgomery despised him. That was one reason. Second, when Oppenheimer was president, the director of the Institute for Advanced Study was Lewis Strauss, who was Jewish and very influential in the government. Oppenheimer caught his attention because he (Oppenheimer) was also Jewish. So, Lewis Strauss took Oppenheimer along. It was in 1947, and Oppenheimer was in California at that time. He went from California to Washington every day as an important figure in the American nuclear policy and defense policy, so he often went to Washington. Lewis Strauss suggested that he come to the Institute for Advanced Study at Princeton so he could just take the train to Washington. Oppenheimer said it was a great idea, but he didn't want to do fund raising, which was and still is the most important thing a university president or head of an institute usually has to do. Lewis Strauss said, "No problem. We have money. We don't need money." So, Oppenheimer went there. After he went to Princeton, prices began skyrocketing, and the Institute for Advanced Study was slowly expanding, so soon there wasn't enough money. Originally, professors at the Institute for Advanced Study were paid 50 percent more than the best professors at other top-notch universities in the United States.

Ji Lizhen: Oh, that's high.

Yang Chen-Ning: For example, Dr. Chern, who went from Chicago to Berkeley, and André Weil, who went from Chicago to the Institute for Advanced Study, all received a 50 percent pay increase. But by the end of the 1950s, it wasn't working anymore. Then the mathematicians at the Institute for Advanced Study were shocked when the institute offered a professorship to Jürgen Moser from New York University (editor's note: in 1970), but he refused. So, the mathematicians at the Institute for Advanced Study said that it had no money anymore. They blamed Oppenheimer and said it was because he hadn't gone out to raise money. You can't blame Oppenheimer at that time because he had said that he wouldn't take charge of raising money when he accepted the position.

Ji Lizhen: Yeah, because they agreed that he wouldn't have to be responsible for raising funds.

Politics at the Institute for Advanced Study at Princeton: A Brief "Love Affair"

Yang Chen-Ning: So, the mathematicians had a story. They said that this situation was like a boy and a girl falling in love. The boy said to the girl, let's get married. The girl said it sounded very good, but she said she couldn't cook. The boy said it didn't matter. But after ten years, it did matter. I heard this story from Dr. Chern, who heard it from some mathematicians. As I said, first they despised Oppenheimer, and second, they didn't know that Oppenheimer had actually made important contributions to physics. They didn't like him going to Washington all the time, and they thought he wasn't raising the needed money, so there were a lot of conflicts, and the most acute conflict was associated with Milnor.

Ji Lizhen: Milnor was from Princeton University.

Milnor and Princeton's Academic Politics

Yang Chen-Ning: Yeah, because Milnor was a professor at Princeton at that time, André Weil suggested inviting Milnor to the Institute for Advanced Study, but Oppenheimer said no. He had a reason for saying this because when the Institute for Advanced Study was just founded, there was no affiliated office, so it borrowed the Fine building,[36] which was the mathematics department building at Princeton. The Institute for Advanced Study then rushed to build its office building. After three years, the construction was finished, and they moved in. Therefore, the Institute for Advanced Study had been housed at Princeton University for several years. There was a gentleman's agreement. At that time, Princeton knew that the Institute for Advanced Study had a lot of money, so they said the institute should not poach people from Princeton. In the late 1950s, when André Weil was going to invite Milnor, Oppenheimer could not deny Milnor's mathematical ability, but he said they had rules, this unwritten verbal agreement, so there were a lot of quarrels about it. At that time, the discussions were very bitter because meetings were held with twenty professors discussing all the important issues concerning the institute. So I sometimes felt like, well … I felt very reluctant to go. But when we went there, our physicists banded together to protect and support Oppenheimer. In the end, André Weil and his fellows failed, so Milnor did not come to the Institute for Advanced Study after all.

Ji Lizhen: He went to MIT first.

Relationship Between Marital Failure and Academic Career

Yang Chen-Ning: But then something else happened. Milnor was getting divorced. In the United States, when a person was about to get

---

[36]The Fine Building (now known as Jones Hall) was named in honor of Henry Burchard Fine, a Princeton mathematics scholar and author of *Fine's Great Algebra*, which was used as a middle school mathematics textbook during the Republic of China (1912–1949).

divorced, other schools knew that this person was likely to leave the school, so MIT invited him.

Ji Lizhen: Oh, so that's the reason.

Yang Chen-Ning: Therefore, he went to MIT. Some years later, when Oppenheimer was no longer the director, Milnor was invited to the Institute for Advanced Study.

Ji Lizhen: Oh, I had heard an incorrect version of this story. I was told that the Institute for Advanced Study wanted to invite him, but they let him go to MIT first and then come back.

Yang Chen-Ning: No.

Ji Lizhen: That's not true, then.

Yang Chen-Ning: It was because André Weil and Montgomery failed, and Milnor happened to be willing to be at the Institute for Advanced Study. I also know his wife very well.

Ji Lizhen: Which one?

Yang Chen-Ning: His wife at that time. Now he has another wife, a mathematician.

Lin Kailiang: She is in *Mathematicians: An Outer View of the Inner World.*[37]

Ji Lizhen: Margaret Dusa McDuff.

Yang Chen-Ning: That's right. His wife at that time was not a mathematician, and Du Zhili knew her well. They got divorced, so MIT invited him.

---

[37]Mariana Cook, *Mathematicians: An Outer View of the Inner World*, translated into Chinese by Lin Kailiang *et al.*, Shanghai Century Publishing Group, 2015.

Yang Chen-Ning and John von Neumann

Ji Lizhen: Well, so that's the reason. I didn't know about that. That's interesting. Also, I want to know if you had contact with (John) von Neumann?

Yang Chen-Ning: Of course.

Ji Lizhen: Because he was later involved in the atomic bomb project.

Yang Chen-Ning: He was a great man, but he died too early.

Ji Lizhen: Yes, he died at a very young age. When I later went to the Institute for Advanced Study, there was another house, and it seemed that von Neumann built the electronic computer ENIAC in that house.

Yang Chen-Ning: (John) von Neumann made the first big electronic modern computer. At the Institute for Advanced Study, there was a small house, and there were a lot of stories.[38] Two of von Neumann's most important fellows helped him. One man responsible for technology named Julian Bigelow. I think you were at Princeton. Didn't you say you were at Princeton in the 1990s?

Ji Lizhen: 1994–1995.

Yang Chen-Ning: I think he was still there at that time. He was about my age.[39] He was an engineer. I think von Neumann himself didn't do any of this stuff but just designed the concept. The real technical things were done by Julian Bigelow. The other man in charge of all this was named Herman Goldstine, so it was these two men working with von Neumann who built it (the electronic computer ENIAC). But neither of them could become a professor at the Institute for Advanced Study because their academic levels were not recognized. It turned

---

[38]Readers who are interested can see George Dyson, Chapters 4–8 in *Turing's Cathedral: The Origins of the Digital Universe*, translated into Chinese by Sheng Yangcan, Zhejiang People's Publishing House, 2015.
[39]Julian Bigelow (1913–2003).

out that when it was all done, Los Alamos Laboratory was in such a hurry to make a hydrogen bomb that the computer was copied before it was even put into use. Thus, the one in the Institute for Advanced Study was called JOHNNIAC (editor's note: acronym for John v. Neumann Numerical Integrator and Automatic Computer) in honor of von Neumann, and the copied one was named MANIAC (editor's note: acronym for Mathematical Analyzer, Numerical Integrator, and Computer). That calculation proved that Edward Teller's original idea couldn't work, so Teller tried something else, and the result was an idea called "radiative compression."

Ji Lizhen: What about Stanislaw Ulam's contribution? He was part of the team. How about him?

Yang Chen-Ning and Ulam

Yang Chen-Ning: I know Ulam very well, too. I think this was the situation; quite like what I just said, it didn't work. The original idea didn't work. When Ulam and von Neumann ran it on the computer,

Yang Chen-Ning and Ulam

it showed that Teller's original idea couldn't work, so he had to find another way. I think the following is the story that Ulam told his wife later, so it's believable. One morning, at breakfast, Ulam said to his wife that "it needs be compressed twice." The principle of a hydrogen bomb is that it uses fusion material that needs to be compressed, which then causes it to become denser and hotter. If the density is big enough and the temperature is high enough, the fusion reaction will start.

Ji Lizhen: I see.

Yang Chen-Ning: The result turned out to be his original method. What was the original method? It was an oval uranium shell with two foci inside it. One focal point was fusion material, and the other was the atomic bomb. When the atomic bomb exploded, its shock wave focused on the other focus and the other focus compressed it. When calculating, although it was compressed, the temperature and density were not high enough, so it didn't work. The sudden thought that Ulam had that morning was to compress it again.

Ji Lizhen: Just one more compression; then his contribution is quite big.

Yang Chen-Ning: But Ulam was not a physicist. He didn't know how to compress it again. When he walked up to Los Alamos Laboratory, he saw Teller and asked him if he should compress it again. Teller was my teacher, and he had many ideas. He came up with an idea to put a kind of foam here, that is …

Ji Lizhen: Foam?

Hydrogen Bombs: Teller and Ulam

Yang Chen-Ning: It's something very light, but it contains a lot of X-ray-absorbing molecules. When an atom bomb explodes, it sends out a shock wave. Its main energy is not in the shock wave, but rather is radiating out and is usually wasted or heats up its surroundings.

Teller's idea was that when this X-ray came out, it would vaporize all the X-ray-absorbing elements in the foam. Once it was vaporized, the pressure would of course be high, so it would be compressed. This compression would be quick. It would be an immediate compression, and then when the shock wave follows, the second compression would occur. The key for this to work is that the time interval between the two compressions has to be just right to have resonance. The idea couldn't have been Ulam's, because he didn't know physics, so the idea was Teller's. Teller went to find one of his graduate students, Frederic de Hoffmann. Hoffmann calculated the size and the time interval to have resonance and wrote a report. Teller had planned to put the names of three people on it, but Hoffman refused. He said the idea wasn't his, and he had just made a calculation, so there were just the two names of Ulam and Teller. If you ask me, I think the story I just told is more appropriate. The reason Ulam couldn't figure out how to do the second compression was because he didn't know anything about physics. Ulam is a strange man. Have you read his biography?

Ji Lizhen: Yes, we were going to translate it.

Yang Chen-Ning: Is it written by Gian-Carlo Rota?

Ji Lizhen: No, it's his autobiography.

Yang Chen-Ning: No, I'm not talking about Ulam's autobiography. Reading his autobiography is not enough; you must read the one that was written by Rota. Do you know this guy?

Ji Lizhen: Yeah, I know Rota from MIT. Do you mean Rota wrote a biography of Ulam?

Yang Chen-Ning: No, Rota has a book, Rota's collection, and one of the articles is about Ulam.

Ji Lizhen: Yeah, I see. Is it called *Indiscrete Thoughts*?[40]

---

[40]Gian-Carlo Rota, *Indiscrete Thoughts*, Birkhäuser Boston, 1996. It was also included in Rota's another book, *Discrete Thoughts*, and the title was "Ulam."

Yang Chen-Ning: Maybe, that sounds similar! If you want to know about Ulam, read this book.

Ji Lizhen: I know the book.

Yang Chen-Ning: One of the articles is about Ulam. If you read it, you will know that Ulam is a very smart man. However, in his later years, he fell ill and could not do specific work after that. When I saw him, he was an interesting person. He started by asking you a question, which might be set theory or combinatorial theory or even poker. And then when you thought about it and tried to discuss it with him, he would lose interest. He was only interested in …

Ji Lizhen: Is his attention span relatively short?

Yang Chen-Ning: No, he was just performing that he knew a lot of interesting things.

Ji Lizhen: It's a little showy, isn't it?

Wu Fan: He seems to be a magician, too. He can juggle.

Ji Lizhen: I don't know about that. I read his biography a long time ago. I remember clearly that when he went to Harvard, he calculated whether it was cheaper to buy a car or take a taxi, and his conclusion was that it would be cheaper to take a taxi.

Yang Chen-Ning: He is a very interesting person, even a little strange. If you read Rota's article about Ulam, you will see that Rota said after he fell ill, he could not even do multiplication.

Landau's Greatness and Tyranny

Ji Lizhen: That was a huge loss of ability. It seems that Landau was the same.

Yang Chen-Ning: Do you mean the physicist Lev Davidovich Landau?

Ji Lizhen: Yes, the physicist Landau, Landau–Lifshitz, from the Soviet Union.

Yang Chen-Ning: Landau is a very good physicist.

Ji Lizhen: I'm just saying Landau used to …

Yang Chen-Ning: No, no, I know the story you're telling. The story is that Landau had a car crash, and after the car crash, while he was in the hospital, someone saw him and said to him, "I heard you can't do physics anymore," and he said, "That's right, but I'm still better than Lifshitz."

Ji Lizhen: Yes, Landau is very interesting.

Yang Chen-Ning: If you want to know the story of Landau, there is a book written by his students.

Ji Lizhen: What's that?

Yang Chen-Ning: It's called …[41] I can't recall the name. I will email you the name of the book next time. It's an interesting book, a collection of articles written by many students of Landau. It's interesting that when you look at these articles, you can see how Landau trained his students. It's very special. No Chinese, French, Germans, or Americans would do it. He asked the students to speak, and he laughed at him and said, "Didn't your mother teach you that when you were a kid?"

Ji Lizhen: It seems to be a Soviet tradition. Israel M. Gelfand was also like this.

Yang Chen-Ning: Gelfand is a bit like Landau, but Landau is even

____

[41]I. M. Khalatnikov, ed., *Landau: The Physicist and the Man. Recollections of L. D. Landau*, translated by J. B. Sykes, Pergamon Press, 1989.

more ruthless with his students than Gelfand. Landau's students all admired him because he took care of them. When students graduated, he helped them find jobs. He had a notebook, and even though he had dozens of graduate students in his life, he wrote down their names and where they went in this notebook like a biographical sketch.

Mark Kac and Yang Chen-Ning

Ji Lizhen: Wrote them all down. That's good. And there's Mark Kac, who does probability. Do you know him?

Yang Chen-Ning: Very well.

Ji Lizhen: Tell us about him. He seems really interesting, and we're translating one of his autobiographies.

Yang Chen-Ning: Does he have an autobiography? I know him very well. In the early 1950s, he visited the Institute for Advanced Study at

Photo of Yang Chen-Ning and Kac

Princeton, and in the late 1970s, he became a professor at Rockefeller University. He was interested in statistical mechanics during those two periods, so we had a lot of disputes about various topics. I only saw him once after he moved to Caltech at a celebration for James W. Mayer in 1982. In my *Selected Papers of Chen-Ning Yang II: With Commentaries*, there is a photo of him. In fact, I have a very famous theorem called the unit circle theorem. Do you know it?

Ji Lizhen: I don't know that one. Tell us what kind of theorem it is, please.

Yang Chen-Ning: The unit circle theorem is about a class of polynomials, which are used in physics. Lee Tsung-Dao and I wrote an article in which we said that the roots of these polynomials are on the unit circle, and this is the unit circle theorem.[42]

Ji Lizhen: What kind of polynomial is that special?

Yang Chen-Ning: This actually came about by chance. This proof is related to Kac in some way. I said in the beginning of one of my articles that we were not able to prove this theorem. It's a set of polynomials, which could be very large or very small in order, and it's a statistical mechanics problem. If it's a three-body problem, it's a three-order polynomial, and it didn't matter how many dimensions it had. Then we said that the N roots of this N-order polynomial were all on the unit circle. We guessed it at first. How did we guess it? If you do a simple two-particle problem first, you can see that three is neither trivial nor too difficult. After doing the four or five, the results are the same. So, we say that this must be a general result. Then we tried to prove it, and it took a long time. We also went to von Neumann and

---

[42]Once Dr. Yang in a speech said, "The reason I could think of considering polynomial zeros is that when I was young, my father taught me two beautiful mathematical theorems, one is a complex coefficients polynomial that can only decompose into the product of a polynomial (and the other one is regular, with 17 sides and can be drawn with ruler and compasses, and it happened to be closely related to the symmetry). It tells us that polynomials are determined by their zeros, so it occurred to me at that time that it was natural to think about zeros. This theorem might be called the fundamental theorem of algebra (historically, Gauss gave rigorous proof of this in his dissertation)."

maybe Atle Selberg. I remember von Neumann said that there was a sufficient condition for this zero-point distribution of polynomials on the unit circle, and Hardy had a book that we could go and find. So, we went to find the book, and found that Hardy said there was a sufficient condition for all the roots of a polynomial to lie on the unit circle, and what was that? It was a set of inequalities. But his theorem was that for an N-order polynomial, there needed to be N inequalities, so this theorem didn't work for us. How could we go on and prove so many inequalities? Then Kac came along, and he was interested in this problem because he was doing statistical mechanics at that time. He said, "I can't prove your theorem, but if you have N particles interacting equally with each other—in our theorem, the interaction is arbitrary—the situation becomes very simple, and I can prove that." So then one day I thought, this was a great idea, and I wondered how I could generalize it. The result was figured out and so was the proof. This is what I still believe to be very beautiful, and it was you (Lin Kailiang) who told me it's in David Ruelle's book …

Lin Kailiang: The David Ruelle's book I gave you has a chapter on this theorem, and the book is called *The Mathematician's Brain*.[43]

Ji Lizhen: OK, I will have a look.

Yang Chen-Ning: Kac has another important contribution. In statistical mechanics, there is a very famous Onsager solution called the Ising model. Lars Onsager was a Norwegian who won the Nobel Prize in Chemistry, but his solution to the Ising model is regarded as a landmark contribution to statistical mechanics. He solved the Ising model in 1944. After that, people became interested in this model. The unit circle theorem I've been talking about is all about that sort of thing. But suddenly, Kac and John Clive Ward put forward a new way of getting Onsager's solution. Ward is a physicist about my age, and he is very smart. Kac and Ward came up with a new way of getting Onsager's solution in 1952 without so many algebra calculations.

---

[43]David Ruelle, *The Mathematician's Brain: A Personal Tour Through the Essentials of Mathematics and Some of the Great Minds Behind Them*, translated into Chinese by Lin Kailiang, Wang Jing and Zhang Haitao, Shanghai Century Publishing Group, 2015.

Onsager's original solution was impossible to understand because it was too complicated, even after simplification. This solution that Kac and Ward developed is a very interesting thing, but it has not been important so far because there have been no new developments. However, anyone who sees this will know that it is wonderful.

Feynman–Kac Formula

Ji Lizhen: How did the Feynman–Kac Formula come about?

Yang Chen-Ning: That was done by him and Feynman …

Ji Lizhen: Was Feynman also there?

Yang Chen-Ning: Kac is a very smart guy. He's a short guy who likes telling stories. Later, he went to Caltech to stay with Feynman, but mathematics at Caltech had never been good. I don't know what is going on now.

Feynman and Yang Chen-Ning, Spring 1957, University of Rochester

Ji Lizhen: It's still not so good.

Yang Chen-Ning: That's because they're being run by a guy named Barry Simon, who's doing applied mathematics, so he hasn't come up with anything important. From 1995 to 2000 when Barry Simon taught calculus to freshmen at Caltech, he said that "ε and δ were the sharpest weapons for calculus students!" One year later, in his last class, the students presented him with a pair of boxing gloves with ε and δ printed on them.

Feynman and His Influence on Caltech

Ji Lizhen: Yes, I was told that the reason why Caltech was not good at mathematics was that Feynman seemed to despise mathematics. Is it true?

Yang Chen-Ning: Yes.

Ji Lizhen: Feynman had a big influence on Caltech.

Yang Chen-Ning: Yes, Feynman was a very smart man, but he didn't think mathematics on its own was good. He seemed to feel that "if we need mathematics in physics, we'll do it ourselves."

Ji Lizhen: Have you read Feynman's book *Surely You're Joking, Mr. Feynman!*[44]?

Yang Chen-Ning: Sure. It's a very famous book.

Ji Lizhen: How many of the stories in it are true?

Yang Chen-Ning: What is written about Feynman in the book is the same as he actually is.

Ji Lizhen: Are you serious about that?

______
[44]The Chinese translation is published by Hunan Science and Technology Press.

Yang Chen-Ning: It's true. That's really how it is. I wrote an article comparing Feynman and Schwinger.

Lin Kailiang: Mr. Ji has bought your book *Collection of Dawn*, and it is in the book.[45]

Yang Chen-Ning: Is it in *Collection of Dawn*? But that's in Chinese. I originally wrote it in English. I'll give you the English one.

Ji Lizhen: English version is better. I think you know this mathematician's story very well. Even though we have heard of many mathematicians' names, your perspective is different. How about Selberg? You and Selberg hang out a lot, too, right?

Yang Chen-Ning and Selberg

Yang Chen-Ning: Of course, I know Selberg very well. I met him for the last time in 2006.

Ji Lizhen: Was it in Hong Kong SAR?

Yang Chen-Ning: No, at the Institute for Advanced Study at Princeton. In 2006, I took Weng Fan, my new wife, to the Institute for Advanced Study. It was a Saturday, and nobody was there. I went to the common room for mathematics, and there sat Selberg, but he did not recognize me. I told him I was Yang Chen-Ning, then he quickly recognized me. Later I asked him how his wife was, and he said he had a new wife.

---

[45]The article is titled "Schwinger." Dr. Yang wrote:

> Feynman and Schwinger are two of the great physicists of our time. Each of them has made many profound contributions. They were both born in 1918. But in terms of personality, they are almost polar opposites. I have often thought that one might write a book entitled *Schwinger and Feynman: A Comparative Study*:
>
> Twenty percent impulsive clown, twenty percent professional nonconformist, sixty percent brilliant physicist. Feynman worked almost as hard to become a great performer as he did to become a great physicist.
>
> Being shy, erudite and speaking and writing in beautifully crafted sentences, Schwinger is a paragon of a cultured perfectionist and a quite inward-looking gentleman.

Ji Lizhen: Really?

Yang Chen-Ning: His first wife was Norwegian. He brought her to the U.S. from Norway. In my impression, that photo, did I give you (Lin Kailiang) that photo? He was gone a year after that photo was taken. He was a few years older than me, so I think he would have been ninety-one or ninety-two when he passed.

Ji Lizhen: He's advanced in age.

Yang Chen-Ning: When we first met, they came to our house for parties, and we also went to his house, but I don't remember having any conversation with him. He didn't seem interested in anything.

Ji Lizhen: Last time Borel and I organized an academic event in Hong Kong SAR, I stayed there during summer for two months. One of Selberg's students who was in Hong Kong SAR invited him, and we spent every morning having breakfast and talking. Then he told me two things that I remember very clearly. First, he didn't think he was a professional mathematician. He said he was an amateur, just playing around. He didn't think he had studied mathematics properly.

Selberg and Yang Chen-Ning

Yang Chen-Ning: Who are you talking about?

Selberg Considering Himself an Amateur Mathematician

Ji Lizhen: Selberg. He saw himself as an amateur rather than a professional mathematician because he hadn't studied mathematics formerly and had taught himself. Because his father[46] used to have some books by Srinivasa Ramanujan in his study, he read them, and his two brothers also read them. They were mathematicians, too.

Yang Chen-Ning: Really?

Ji Lizhen: Yes, he just looked around. I shared this with Borel, who laughed and said if Selberg didn't think he was a mathematician, what about anyone else?

Yang Chen-Ning: Do you understand his elementary proof of the prime number theorem?

Ji Lizhen: I don't know. I've only just heard about it.

Yang Chen-Ning: He became famous because of that.

Ji Lizhen: Yeah, and he won the Fields Medal for it.

Yang Chen-Ning: Yes, but you didn't study it?

Dispute Between Paul Erdös and Selberg

Ji Lizhen: No. It was for that reason that he had a quarrel with Paul Erdös, and they had a dispute.[47] Later, Selberg did mathematics

[46]Ole Michael Selberg, Atle Selberg's father, was a mathematician. He had four sons. Of the twins, Sigmund and Arne Selberg, born in 1910, Sigmund became a mathematician and Arne an engineer. His other sons, Henrik Selberg and Atle Selberg, were also mathematicians, born in 1906 and 1917, respectively.
[47]See the Chinese translation of Erdös's biography, Chapter 8, *My Brain is Open: The Mathematical Journeys of Paul Erdös*, translated by Wang Yuan and Li Wenlin, Shanghai Translation Publishing House, 2002.

wonderfully because he started very narrowly. He spent most of his time doing zeta functions, elementary proofs of the prime number theorem, and then the discrete groups in Lie groups, Selberg trace formula and Selberg's hypothesis. You see, Gregori A. Margulis and David Kazhdan both worked on that. So, it's a wonderful change, from a very narrow focus at first to very wide later.

Yang Chen-Ning: I didn't talk much with Selberg, but I can tell you two stories about Harish-Chandra.

Ji Lizhen: Really? That's interesting.

Harish-Chandra's Story

Yang Chen-Ning: One story is that in the early 1960s, when he first came to the Institute for Advanced Study, it happened that Paul Cohen at Stanford had solved the continuum hypothesis. So, one day I ran into Harish-Chandra and said, "This is a big breakthrough. Do you want to change to that field instead?" He said, "It's a field which one can get in but never get out of."[48]

The other story is that he was a bit arrogant. One day he said something strange to me, but certainly I couldn't ask him about it in detail. He suddenly told me that Professor Chern Shiing-Shen was a strange mathematician, and his tone gave me a negative impression.

Ji Lizhen: Really?

Harish-Chandra and Chern Shiing-Shen

Yang Chen-Ning: Harish-Chandra said that Chern Shiing-Shen was a very strange mathematician.

Ji Lizhen: They have different tastes.

---

[48]In an English e-mail from Dr. Yang to one of the editors, Harish-Chandra replied, "No. I do not want to become sterile."

Yang Chen-Ning: Actually, they are not related at all, and I don't think their fields are closely related, so he made the judgment from a distance. If you want me to make a guess, his method of studying mathematics is different from Dr. Chern's; he doesn't know much about Dr. Chern's mathematics, and he's a little jealous and disrespectful, I think.

Ji Lizhen: Yeah, probably, because Harish-Chandra's work is not fully recognized. Because he had been told …

Yang Chen-Ning: Do you know him?

Ji Lizhen: I know about him, but I've never met him.

Yang Chen-Ning: I'm afraid he was not there when you went.

Ji Lizhen: He passed away. He died in 1983 when Borel was 60 years old.

Reasons Why Harish-Chandra Was Not Awarded the Fields Medal

Ji Lizhen: Yeah. Several Indians in our department told us that when Harish-Chandra was nominated for the Fields Medal, Carl Ludwig Siegel was the judge. He thought Harish-Chandra was doing something too abstract, Nicolas Bourbaki style, and he said no. Harish-Chandra may have felt that he had changed the field of representation for half a single Lie group, and might have felt bitter for not winning the prize. You really know a lot of mathematicians, many of whose names I've heard of but never met.

Yang Chen-Ning: Let me say a few more things. In the letter you (Lin Kailiang) sent me about Hua Loo-Keng, a man named Wang Yuan[49] said that Weyl asked Chern to go to the United States with Hua, but for some reason, Chern went there in 1943 while Hua did not.

---

[49]Wang Yuan (1930–2021) was a distinguished Chinese mathematician, renowned for his contributions to the Goldbach Conjecture.

Wang Yuan explained that Hua was doing the same thing with Siegel at that time, and he was afraid that after he went there …

Ji Lizhen: Was he invited by Siegel?

Lin Kailiang: He's afraid that people might say he was Siegel's student.

Yang Chen-Ning: Learned from Siegel. This is what you said in your letter.

Lin Kailiang: Yes, this was the explanation given by Wang Yuan.[50]

Yang Chen-Ning: I think this issue is worth studying. You should do it while Wang Yuan is still alive. I think the reason is not that simple, and in this case … I think so. Of course, Wang Yuan may not know all the details, but Wang Yuan knows more about this matter than you and I do. It is even possible that Hua Loo-Keng told him about it. Then my impression is that it is a very important job. The question of whether or not to represent this kind of group matrix is probably a new direction, which Weyl, Siegel and Hua Loo-Keng all know; it is very important. So, I think it's worth digging into.

Ji Lizhen: Siegel is an interesting guy.

---

[50] The reason given by Hua Loo-Keng himself was reflected in his letter to the then Minister of Education, Chen Lifu, written on January 15, 1944. In terms of research, Loo-Keng and Siegel, a former professor at Göttingen University, independently developed "the theory of automorphic functions of number matrices," and Loo-Keng's theory is more extensive and sophisticated (which was confirmed by himself and Weyl). As Professor Weyl said, the mine of important results which Loo-Keng obtained is inexhaustible. Siegel is now at the Institute for Advanced Study at Princeton. If Loo-Keng goes there, we will learn from each other. However, his reputation in academia is higher than that of Loo-Keng due to his age and experience. If Loo-Keng (received the Institute for Advanced Study's invitation and) was to go there, Loo-Keng might sacrifice the reputation as an independent inventor and become part of the Siegel school. Loo-Keng might have to go abroad after setting a basis (it would be better if Loo-Keng could go there without being invited by the Institute for Advanced Study). If my research can be seen by the world, it will be one of the best stories in the War of Resistance against JapaneseAggression.

Hua Loo-Keng and Chern Shiing-Shen's Time in Kunming

Yang Chen-Ning: Of course, there is another possibility. At that time, at least in Kunming, people thought that Hua Loo-Keng contributed more than Chern Shiing-Shen.

Ji Lizhen: Yes, he won many prizes. And it seems that there is still much publicity in the museums.[51]

Yang Chen-Ning: And Hua was invited to the Soviet Union and so on. So, there's another possibility, he felt he didn't need to go to the United States. This could be (the reason for his deferred visit to the Institute for Advanced Study).

Lin Kailiang: According to the letters between Hua Loo-Keng and Weyl, one reason is that Hua felt that the salary offered by the Institute for Advanced Study was not enough for him to support his family, and the fund ($1,500 per year) also was not enough.

Yang Chen-Ning: Who paid for his visit to the Soviet Union?

Lin Kailiang: Maybe the Soviet Union invited him.[52]

Yang Chen-Ning: I think the invitation from the Soviet Union has something to do with the country. The invitation from the Institute for

---

[51]See Ji Lizhen, "Singularity in Kunming Time and Space—Southwest Associated University," included in the 12th volume of "Mathematics and Humanities" series *One Hundred Years of Mathematics*, edited by Yau Shing-Tung *et al.*, Higher Education Press, 2014.

[52]According to an article introduced by Dr. Wang Tao of the Institute of History of Natural Sciences, the Chinese Academy of Sciences (article title "Correspondence of Academia Sinica and Others for Hua Loo-Keng's Investigation and Research in the Soviet Union," *Archives of the Republic of China*, Issue 4, 2007), Hua Loo-Keng's visit to the Soviet Union was fought for by himself, supported by Academia Sinica, and facilitated by Mr. Jiang Lifu. From March 1941, Jiang Lifu successively served as a full researcher and director of the Preparatory Department of the Institute of Mathematics of Academia Sinica.

Advanced Studies in the United States was just for the institute. You can ask Wang Yuan.

Yang Chen-Ning's Health Secret

Ji Lizhen: You have such a good memory. You remember things from a long time ago so clearly.

Yang Chen-Ning: I guess I'm lucky because my brothers are much younger than me, but they didn't have as good a memory as I do now.

Ji Lizhen: Yeah, and you look so energetic.

Yang Chen-Ning: My problem is that I have aches and pains here and there. If I don't cover myself properly when I sleep, I always have aches and pains the next day. Now, I can deal with it. The 301 Hospital told me that I can sleep wearing a corset-like thing that foreign women putting on in the past.

Ji Lizhen: Do you usually do any exercise?

Yang Chen-Ning: No.

Ji Lizhen: You don't exercise?

Yang Chen-Ning: I'm too lazy.

Ji Lizhen: You are in good health and good spirits.

Yang Chen-Ning: Well, I told Lin Kailiang that the area of popular science and, for example, biographies of scientists is what China should develop, and I think there are millions of college graduates every year, many of whom I think are majoring in liberal arts. What are they doing out there? So I think this is an area that can encourage people to make progress.

Ji Lizhen: That's good, but there's a problem right now. If you want to do something like that, you need organizations to sponsor them.

Yang Chen-Ning's Biography

Yang Chen-Ning: Yes, that's absolutely true. I don't know if you've read one of my biographies written by Jiang Caijian.[53]

Ji Lizhen: Yes.

Yang Chen-Ning: It was published in Chinese Taiwan.

Ji Lizhen: Yes, I have it. It's very good.

Yang Chen-Ning: The reason why he was able to write that biography was that there was a wealthy newspaper[54] owner in Chinese Taiwan who was a former classmate of Wu Chien-Shiung. When Wu Chien-Shiung got old, the owner came to Jiang Caijian saying that he would give money for his two-year stay in the United States if he wrote the *Biography of Wu Chien-Shiung*. The biography is very good. This biography was written according to the modern Western method of writing a biography of science. After he finished writing the *Biography of Wu Chien-Shiung*, he wrote my biography. There is always someone willing to support his expenditure.

Wang Liping: Now the country is doing this, slowly. The Beijing Science and Technology Association is now encouraging popular science writing.

Yang Chen-Ning: Especially now that the Internet has been set up, there are many problems in the publishing industry.

Ji Lizhen: That's right. I do have an idea. Symmetry is something we all

---

[53]Jiang Caijian, *The Beauty of Norms and Symmetry: Biography of Yang Chen-Ning*, Guangdong Publishing Group, 2011.
[54]*The China Times.*

Yang Chen-Ning Biography cover design, in simplified and traditional Chinese

know, but the general populace only vaguely knows what it is. If you ask a few more questions, they might not know anymore, and your work is very related to symmetry. Could you please talk about symmetry in an easier and in-depth way? Symmetry exists in many unexpected places. For example, the general populace knows the gauge field, and your previous Nobel Prize-winning work is related to symmetry, but they know relatively little about the relationship between the Yang–Baxter equation[55] and symmetry. There are a lot of people coming to Beijing in summer, so I would like to ask experts to inspire them a little bit. This motivates some young people to begin learning. Once they cross the threshold, they can "practice by themselves."

---

[55]On February 20, 2016, BBC Earth (formerly known as BBC Knowledge, a documentary and information channel owned by the British Broadcasting Corporation) published a list of 12 alternative options for the most beautiful mathematical formulas, among which was the Yang–Baxter equation. See http://www.bbc.com/earth/story/20160120-you-decide-what-is-the-most-beautiful-equation-ever-written.

The Second Interview

Time: March 1, 2017
Place: Institute for Advanced Study at Tsinghua University
Interviewer: Ji Lizhen, Wang Liping
Recorder: Wang Liping
Collator: Peng Cheng, Wang Liping, Ji Lizhen

Before we went to interview Dr. Yang, we prepared a long list of questions, but the topics of our interview were quickly derailed by Dr. Yang. He diverged in the directions at the pace that interested him, thus covering many unplanned topics.

- There was a lot of talk about the marriage between Yang Chen-Ning and Weng Fan and how Dr. Yang viewed the age difference between them.
- Yang Chen-Ning is of course best known for the Yang–Mills theory, the non-Abelian gauge field theory. How does this relate to Weyl's Abelian gauge field theory, and why did Weyl miss it?
- What does Yang Chen-Ning think of the many physicists and mathematicians, such as Einstein and Feynman?
- How should we judge the contributions and influence of von Neumann, Dyson, Gelfand *et al.*?
- How are Nobel Prizes related to politics? Who should get them?
- What's the difference between a physicist and a mathematician?
- What is the non-academic life of a great man like? What is the relationship between them (for example, Feynman, Dyson, Ulam and von Neumann)?

You will get answers to the above questions in this interview.

What Is Youth?

Ji Lizhen: Over the course of history, only a few individuals truly understand things and have an impact. I think you're one of the most important people among them. I've thought about it, and there are at least five reasons. First of all, you've done really great work, and it's work that has a long-term impact, not just for one moment in time. Among the leaders in physics and mathematics and other important people in all aspects of government, you've seen a lot and you're still learning. Recently I came across a poem called "What Is Youth?" written by a little-known American. Later, an American general in Japan, named MacArthur, kept the poem in his office. The poet became famous in Japan because he encouraged the Japanese to rebuild their country after World War II. Then in the 1980s, I think, a Japanese man found the poet. (Information about the poet and the poem is in the preface.)

Yang Chen-Ning: To find this American poet?

Ji Lizhen: Well, yes, to find this American poet. The Japanese man found his house, bought it back and donated money to build a museum because the Japanese felt that the poet had inspired a generation of people. The English version of the poem is in this journal, and I recently wrote an article about that event and the poem.

Yang Chen-Ning: Let me see. Is this his poem?

Ji Lizhen: This is the journal you asked for. It's called *Mathematics, Science, History and Culture*. It's the latest issue.

Yang Chen-Ning: It's this one. I can't get a subscription here because it's a magazine in Chinese Taiwan. Is it issued four times a year?

Ji Lizhen: Yes, my article is the last one. We can look at the beginning of the article. I think you are very young by his definition because he said that a person's youth was not defined by age but mentality.

Look at his first words. I liked this poem so much that I wrote a few sentences later.

Yang Chen-Ning: Do you have the original English version of it?

Ji Lizhen: Yes, the English version is here and the Chinese one is there.

Yang Chen-Ning: Did you translate the Chinese version?

Ji Lizhen: I didn't translate it. I can't type Chinese.

Yang Chen-Ning: Who did it?

Ji Lizhen: There are many different versions on the Internet.

Yang Chen-Ning: You know, when I married Weng Fan in 2004, there was a lot of criticism because we were so different in age. There were criticisms then, and there are still criticisms now. But when I got engaged to Weng Fan, I wrote a letter to my siblings, my children and some close friends. There is a saying in the middle of the letter which states that youth is not entirely determined by age. I will remember to email a copy of that letter later.

Ji Lizhen: Okay, I will have a look.

Yang Chen-Ning: Great, thank you. Can you leave this magazine for me?

The Majority Is Always Wrong

Ji Lizhen: Sure. We'll read that poem later. Besides, as I said, there were only so few important people in this world. Socrates said thousands of years ago that "the majority is always wrong." I can use democracy as a good example of what Socrates said. Usually, democracy is the majority rule. The Western world is always proud of its democracy. Now, a prime example that carries negative consequences is the

U.S. presidential election, where the democratic reasons lead to the selection of such a president (Trump). Another example, based on my personal feeling, is that I don't own a smartphone and I am not heavily reliant on the Internet, but I believe that there is a terrible phenomenon happening now. Everyone has a phone in his hand, so he can express the opinions at any time, and sometimes the opinions are bad or false, and then everyone passes them around. I think this kind of person generally shares bad information and then embellishes it. I think the first one is wrong most of the time, and the second one is even less believable. For example, many of the statements about you on the Internet are not so good. This is my personal opinion. There's a very important Morse theory in mathematics. Have you heard of it? It's how you determine the topology of a manifold, given a manifold. Given a Morse function on a manifold, the information of all topologies lies in the critical points. In a sense, this reflects the structure of a society as a whole: there are a bunch of people who are unimportant and don't know anything. So that's why I think you know so much more than most of us, from the work you've done in academics to the things you've experienced.

I want to talk to you about some questions this time. The first goal is to listen to your comments on some questions that are of interest to most of us. Basically, I will organize it in this way. For students and scholars, I think there are several steps in their process: firstly they read, do research, develop their careers, and then live their lives. Since scientists are human beings after all and need to have their own lives, I think you can give some principled advice and opinions on this aspect. I forgot one question just now. I think you have had some contact with H. M. Morse when you were at Princeton, haven't you?

Morse and the First Application of Topology in Theoretical Physics

Yang Chen-Ning: Yes, but not much because he is a bit older. When I went to Princeton in 1949, he was at Harvard at the time, and then he came in the 1950s. When he came, there were seven or eight mathematicians and six or seven professors at the Institute for Advanced Study, and he was also one of the older ones among them.

He and his wife sometimes invited us to their house for tea and so on. So, he was very different from von Neumann. He was much younger and very different from Selberg and Montgomery, and this theory of his, I hear, is very important in mathematics.

Ji Lizhen: It's very important.

Yang Chen-Ning: Because he was one of the pioneers of topology. Did I tell you that the first application of topology in theoretical physics had something to do with Morse?

Ji Lizhen: I don't think so. Can you explain that again?

Yang Chen-Ning: Give me a pen, I'll email you an article about the Morse article.

Ji Lizhen: Okay. I just recalled, the topology of a certain manifold is a bit like the Morse theory study. Because most people really don't understand it and why it matters, they just go with the flow. Ordinary people don't have any idea about it, and that's why I think some of your ideas are particularly important.

Estimated Number of Chinese Mathematicians with Doctorate Degrees Abroad

Yang Chen-Ning: I also have a question for you. Since the Chinese economic reform, there must be hundreds of Chinese mathematicians who have gone abroad to get their Ph.D. degrees.

Ji Lizhen: Yes.

Yang Chen-Ning: Do you have an estimated number? I think the number is less than one thousand.

Ji Lizhen: I'm not sure.

Yang Chen-Ning: Has no one studied it either?

Ji Lizhen: No one has studied it.

Yang Chen-Ning: As the reform has been carried out for more than 30 years, I think there must be at least 300 people.

Ji Lizhen: Definitely. A better way to get statistics is to find out how many Chinese professors there are in the mathematics department of every American university. If you include those majoring in statistical probability, there will be even more.

Yang Chen-Ning: Do you think there are more than 10 a year in the United States?

Ji Lizhen: You mean those who got a Ph.D.?

Yang Chen-Ning: Chinese people who got a Ph.D. in the United States.

Ji Lizhen: More than that. There are at least 50 Chinese people who get their Ph.D. every year, but most of them can't stay there. There are a lot of people who get their Ph.D. in America.

Yang Chen-Ning: Do most of them do other things later?

Ji Lizhen: Yeah, more than 50.

Yang Chen-Ning: So, there's no problem with several hundred, is there?

Ji Lizhen: No problem.

Comparison of Chinese Mathematics and European Mathematics

Yang Chen-Ning: Do you feel that, compared with many countries in Europe, the results are not satisfactory?

Ji Lizhen: Not satisfactory. I can give two examples. Israel is a small country, but there are a lot of good mathematicians. France, too. Being good is not a matter of numbers.

Yang Chen-Ning: Did you discuss the reason? And the Russians. The Russians are also good at this.

Ji Lizhen: The Russians are not good now. They're not as good as they used to be.

Yang Chen-Ning: Do you mean young Russians today?

Ji Lizhen: Russian mathematics itself is far from what it used to be. That's beyond comparison.

Yang Chen-Ning: Why is it not comparable? Is it because the teachers are gone?

Ji Lizhen: Because the environment has been lost. I've been told that, in the past, for political reasons, especially for Jews, the only thing they could study was mathematics or something very theoretical. It was the only way out. Now they can study other things.

Faddeev and Feynman Diagram

Yang Chen-Ning: Do you know that an important 82-year-old Russian mathematician named Ludwig D. Faddeev has just died? Faddeev is mainly in the field of physics. He should have won the Nobel Prize, and people can't understand why he did not.

Ji Lizhen: Yes, I know him. Faddeev. I invited him to Sanya Mathematics Center for a meeting, but he didn't show up.

Yang Chen-Ning: He was in St. Petersburg. He made a lot of contributions, from those everyone thought that he deserved a Nobel Prize. What were his contributions? This can be stated in a nutshell.

You know field theory, which started in the 1930s when you put electromagnetism and quantum mechanics together. But by the end of the 1940s, it was so hard for field theory to calculate the probability of anything. However, someone wrote a book explaining how to do it, and this guy was Feynman, and he figured out how to do it.

Ji Lizhen: Did you say the Feynman diagram?

Yang Chen-Ning: It's the Feynman diagram. He didn't know how to prove it. He guessed it out. He drew a diagram, like this, and then he wrote down every basic part of it. He put it all together, and it became a formula, and the formula was the result of the calculation. But he couldn't prove it. He couldn't tell you how he proved it. In other words, he did not derive it from field theory; he just guessed it. It was later proved by Dyson that the thing he guessed could be derived from field theory. So, Dyson should have won the Nobel Prize, but Dyson didn't. After that, Mills and I wrote the non-Abelian gauge field theory. Because the electromagnetic field is an Abelian gauge field, and the non-Abelian picture is more complicated than that. For example, if you have four at one point, there are only three forces at each point in the electromagnetic field, so it's much more complicated. By the 1960s, everyone was trying to figure out what a Feynman diagram was for a non-Abelian gauge field theory, and Feynman himself was also trying to figure it out. However, the man who figured it out was Faddeev.

Ji Lizhen: A wonderful man.

Yang Chen-Ning: He's amazing, so people think he deserves a Nobel Prize. There is still an unresolved problem in this case. One of his students studied the impact of Faddeev's work in this area.

Ji Lizhen: Faddeev won the Wolf Prize later, didn't he?

Yang Chen-Ning: No, he won the Shaw Prize, along with Arnold. One year, the Shaw Prize was given to the two of them.

Vladimir I. Arnold

Ji Lizhen: Arnold is great, too. Do you know Arnold? I've never met him.

Yang Chen-Ning: Arnold is very interested in history. It seems that you translated a book about it, isn't it?

Ji Lizhen: Yes, the Higher Education Press published his book. Arnold is interesting anyway. After the Fields Medal was announced in 1994, he wrote a letter strongly objecting to the results. Some people said it was because one of his favorites didn't get selected.

Yang Chen-Ning: Arnold is very smart.

Ji Lizhen: His work is very original, and he is a very thoughtful person. You were talking about the Yang–Mills theory. I have a question for you later. I met someone at the meeting yesterday. He was a student of Michael F. Atiyah's student. When he heard that I was coming to interview you, he made me ask you a question. Why? Because when he was interviewing Atiyah, and Atiyah was talking to him about you, one of the questions was a little weird, so we'll talk about it later.

Yang Chen-Ning: What's the name of this person?

Ji Lizhen: It's Oscar Garcia-Prada, a student of Nigel Hitchin, and he interviewed Atiyah.

Yang Chen-Ning: When was the interview, the most recent?

Ji Lizhen: It just came out last year.

Yang Chen-Ning: European Mathematical Society, isn't it?

Ji Lizhen: Yes, it just came out. The interview was conducted a little earlier.

Yang Chen-Ning: You can email it to me later.

Ji Lizhen: I'll email it to you. I'll show you what's in it later. I've highlighted all the issues they talked about. You may read this question about you now.

Yang Chen-Ning and Weyl at the Institute for Advanced Study at Princeton

Yang Chen-Ning: A lot of people have asked me about this because I wrote the article with Mills in 1954. Weyl is …

Ji Lizhen: He died in 1955.

Yang Chen-Ning: I think he died in 1956.

Ji Lizhen: The interview says that it was 1955.

Yang Chen-Ning: 1955, okay. I was in Brookhaven in 1954, and I went back to the Institute for Advanced Study in the fall of 1954, and at that time, Weyl was traveling back and forth between Princeton and Zurich for most of the time. Usually, when he was at Princeton, people would eat lunch in the cafeteria (the cafeteria cooked the best meals at Princeton), but people doing physics and those doing mathematics didn't usually sit together. Weyl was older, and I didn't see him eating in the cafeteria. Oppenheimer had cocktail parties at his house, and Weyl always went there, so I'd seen him a lot of times. Most of the times I met him were at these cocktail parties.

Ji Lizhen: Were these social ones?

Yang Chen-Ning: No, they were always cocktail parties at Oppenheimer's house. He liked to talk, too. I remember him asking me all the time if I knew Hua Loo-Keng and Chern Shiing-Shen. I once wrote an article about Weyl. Have you read this article?

Ji Lizhen: Yes, it's his memorial.

Yang Chen-Ning: It's the 100th anniversary of his birth.

Ji Lizhen: Yes, the 100th year. There's also Borel's article in it. I have this.

Yang Chen-Ning: At the end of the article, I said that he was still working on the gauge field theory in his later years. He developed the theory from 1919 to 1929, and in the 1950s, he was still obsessed with it, so he paid a lot of attention to it. But the gauge field theory that he was interested in was Abel's. If you had asked him whether he wanted to extend his Abelian gauge field theory to non-Abelian gauge field theory, he would have been very happy, and he would have done it very quickly. Somehow, even though Oppenheimer, Pauli, and I all had contact with him, none of us mentioned this.

Ji Lizhen: And you didn't tell him about it, did you? Not when you were doing it.

Reasons Why Weyl Failed to Discover Non-Abelian Gauge Field Theory

Yang Chen-Ning: You have read this article in his later years, and my article on Weyl cited it. After this article, wherever he was, he paid great attention to the gauge field theory. So, from the point of view of his mathematical philosophy, his gauge field theory was all in one place, which was, of course, his main point, but he didn't make it non-Abelian. I talked about him for a long time, and I said that gauge fields and non-Abelian gauge fields were the most interesting things in his life, and yet, he hadn't connected the two.

Ji Lizhen: Yeah, that's why Atiyah said that.

Yang Chen-Ning: So, Atiyah and a lot of people wondered how that could be possible. The result was what Mills and I came up with from the physical direction.

Ji Lizhen: You thought the language and the idea were different, so you didn't discuss it.

Yang Chen-Ning: Because unlike him, I didn't talk about mathematics. I didn't talk about physics. I just made some unrelated conversations at cocktail parties.

Ji Lizhen: And you didn't talk about that in regular conversations as well, did you?

Reasons Why Yang Chen-Ning Discovered Non-Abelian Gauge Field Theory

Yang Chen-Ning: At that time, I think he probably had some discussions with young mathematicians. I don't think he ever had any discussions with physicists.

Then of course there's the question of why Oppenheimer or Pauli didn't tell him. Part of the reason was that Pauli, Oppenheimer and I all knew that the non-Abelian gauge field theory needed to produce massless particles, and there were no massless particles in physics, so it didn't work. Pauli never published papers about it, even though he was also working on this stuff.

But I think I got one thing right, and I know I'm not going to solve the problem, which is why there are no massless particles. But I thought it was so wonderful that I wrote about it. *Physics Review* accepted this paper very quickly. Even without asking this question, it was published within a month.

Ji Lizhen: He (Oscar Garcia-Prada) thought it was wonderful too. That's a good answer. I'll tell him when I see him tomorrow because he told me to ask about it.

The Importance of Fiber Bundles in Physics

Yang Chen-Ning: This happened in the 1950s, and in the 1960s, there's another story. One day, when I was teaching general relativity, which naturally used Riemann geometry and Riemann curves, I wrote the formula on the blackboard. I remember it very well, and I wrote it on the blackboard as it was a long formula. Upon writing, I said that this was very much like the paper Mills and I wrote as some contents in the front part of the paper were linear and some in the back were quadratic, and this was very similar to that. So, I went back to my office after class and compared the two. When I figured it out, I realized they weren't just somewhat alike; they are exactly the same, except that the group was a Poincaré group, not a SU(2). That's when you took this SU(2) and turned it into a more complex group, especially a Poincaré group, it turned into that. So, I went over to James Harris Simons, who was the head of the mathematics department. Simons read it, and he said that was not unusual that both were fibrous bundles. That was the first time I heard of fibrous bundles. Then I asked what fiber bundles were. He gave me a book by Norman Steenrod called *The Topology of Fiber Bundles*. I read it, and I couldn't read it through because it wrote each theorem followed up with a proof. Later, I went to Simons and asked him to explain it to us physics guys in simple mathematical language, and he explained it to us. We had lunch every day, almost formally, and bombarded him with questions. Two weeks later, I finally got it. After I understood it, I wrote an article with Wu Dajun, which, in fact, seemed very simple to mathematicians, but it had its own importance based on which we compiled a dictionary, a very small dictionary.

Ji Lizhen: Yeah, the Yang–Mills theory and the…?

Yang Chen-Ning: In the Wu–Yang dictionary that I compiled with Wu Dajun, On the left side are terms related to physics, and on the right side are nouns related to fiber bundles. In fact, many mathematicians have thought that this is the same thing, but they did not express it precisely. And for the first time, we pointed out that Paul A. M. Dirac had a magnetic monopole called a Dirac monopole, which was a nontrivial bundle. I remember showing Simons the paper by Dirac, and he said actually Dirac had discovered nontrivial bundles decades before mathematicians. So, we perceived this nontrivial bundle an item. Coincidently, Isadore M. Singer came over, and I gave him a copy of this preprint. He took it to Atiyah at Oxford. He and Atiyah studied it with Nigel Hitchin. There seems to be one other person, too. There seem to be four.

Ji Lizhen: Someone from the Soviet Union.

Yang Chen-Ning: All three of them. I don't know if there's the fourth person.

Ji Lizhen: Is it Vladimir Drinfeld and Yuri Ivanovich Manin?

Yang Chen-Ning: Yes, it's them, using Atiyah–Singer theory and stating how many parameters the solution of instantons had. As soon as their paper was published, mathematicians became interested in gauge field theory.

How to Read and Learn

Ji Lizhen: Hua Loo-Keng proposed a method of reading in the preface of his famous book *Introduction to Number Theory*. He said that you have to read the book thoroughly first and then read it in a condensed way. My Chinese is not good (literally translated as reading a book "from thin to thick, from thick to thin"). What's your advice on how to read a book? You probably have also read a lot of books and seen a lot of things. What do you think are the good ways to read?

Yang Chen-Ning: I think it depends. I think different people have different ways of reading.

I don't read many articles or books on physics and mathematics. For example, when I hear that something is developing, my approach is usually to think about it for myself, rather than reading other people's articles or reviews. Sometimes, there are some results, sometimes there is no result, but no result is not a waste of time, because it is different from not thinking for three days and reading the article. I think some people are not into this way. In other words, I do not like to vaguely know a thing, but some people feel okay to know things in a very vague way. So, there are two ways of learning: one is the immersive way of learning, and I have a name for it.

Ji Lizhen: Does immersion mean thorough understanding?

Yang Chen-Ning: The immersive way of learning is to understand vaguely.

Ji Lizhen: It's just a little bit of immersion.

Yang Chen-Ning: You don't truly understand it, but the second time you listen to someone talk about it, you find that you understand it a little bit more than last time. It's a way of learning.

I think everyone can learn with two types of learning methods, namely immersive and accurate learning methods. But the proportion is not the same. I am more focused on the accurate learning method, but I know that the immersive learning method is very important. However, Chinese students, influenced by their learning attitude in middle schools and universities, do not like immersive learning. I had once told a story. There were a lot of students from Chinese mainland who came to Stony Brook University, and I told them that every week the department would invite someone to make a comprehensive colloquium. When I suggested they attend the colloquiums, they said they went there but couldn't understand it. I said I didn't understand it either. But if you did not understand a talk once, and you thought about it later, you would understand it a little more the second time. This method is called the immersive learning method. I told my Chinese students that people trained by the Chinese educational method should pay special attention to the immersive method of study, as it is rejected by the traditional Chinese learning method, but it is an important learning method.

Ji Lizhen: This is related to the other question I've prepared here. Many people criticize that Chinese learning methods are rote learning, and that Chinese students are not quite creative. A lot of people say that middle school students do well in exams but they don't have many ideas nor much originality. I think you were educated both in China and in the West. I remember you wrote somewhere that you benefited from staying in both places and received a better education on both sides. Now my question is, can you tell me the advantages and disadvantages of Chinese education and Western education?

Comparison of Chinese Education and Western Education

Yang Chen-Ning: This is a very important topic. Chinese education and Western education have their own advantages.

China pays too much attention to discipline, so the students are less daring, which weakens their innovation. I think this statement makes sense, but we can't say that Chinese education is absolutely bad. Well, when I think of the best students who earn grades over 90, the American way is better. Why? Because American students are not disciplined all the time. They are unbounded and has unlimited possibilities. If they don't know much about something, they will think about it for three days or talk to someone, and then they will fill the knowledge gap. But students with grades of 80 or so can't do this thing, and the American way is not good for this cohort. I think the first point is that for students with grades over 90, the American way is better; however, for students with 80 or so, the Chinese way is better. The second point is that people who are trained in China should realize that they have this weakness, so they should be braver.

Ji Lizhen: I know that to really make something great and impactful, you have to have something original because repeating someone else's stuff doesn't necessarily have much impact. The question is, you said just now that being bold helps in becoming innovative, but what are the things, or what are the ways in which people can become more creative? This is the goal, so how to attain it?

Yang Chen-Ning: I can share my own experience. When I went to the University of Chicago, I immediately became the best theoretical graduate student in the department. Why? Because when I was in

Kunming, at Southwest Associated University, the most important thing was that I learned both areas very well. One was field theory, which I learned from a man named Ma Shijun, who is not well known in China now.

Ji Lizhen: I haven't heard of him.

Yang Chen-Ning: I think someone should write a biography of Ma Shijun. Ma Shijun was born in 1913 or 1914. He was a bachelor's student at Peking University, and then he went to England to get a doctoral degree. His thesis in England was on field theory. At that time, there was no Feynman diagram in field theory, and there were many people working on it. When he returned to Kunming, he opened a course on field theory. There were very few people who took field theory. I remember there were only three or five persons. He taught it in great detail. I knew a lot about it. So, when I got to Chicago, it wasn't just my group of graduate students who weren't as knowledgeable as I was. Not only did I know a lot about field theory, but I absorbed his spirit and was far better than my fellow graduate students. The best and most important theoretical physicist of the time was Fermi. Fermi was a great contributor to field theory, but he hadn't worked on it for ten years because he was doing experiments. Thus, he didn't know much about field theory in the 1930s and early 1940s. I knew more than he did.

Ji Lizhen: He did not keep up with the times.

Yang Chen-Ning: So, I guess my understanding of the field theory is not just some concrete phenomena, but my understanding of its spirit, which I only understood later.

Ji Lizhen: Do original work.

Yang Chen-Ning's Motivation to Learn Gauge Field Theory

Yang Chen-Ning: Firstly, I had that foundation. And secondly, at that time, there was a new stimulus for experimentation and there was a new field called particle physics. The reason for this was that several experiments were being carried out, and a new cosmographic particle called the τ particle was found, and another particle called the V particle was also found. Within a few years, a number of particles had been discovered, and everyone knew that there was something very wonderful here, something very fundamental, so they wanted to find out what the relationship was between them, and I was doing that, too. But I did not follow others' approach, and I wished to change my original field theory to one that could express these new concepts. Thus came the non-Abelian theory. None of my peer graduate students at that time were going in that direction.

Ji Lizhen: Because they don't have your foundation or your understanding of field theory.

Yang Chen-Ning: Like Dyson. Dyson was very famous at that time, in the early 1950s. I have just mentioned Feynman's diagram which he couldn't prove; Dyson proved it for him, so Dyson immediately became the most famous person in field theory. What research methods should be used for the new particles? Dyson did some research. I'll send you an article later on.

Ji Lizhen: The article about Dyson?

Freeman Dyson's Exploration of Field Theory

Yang Chen-Ning: It was Dyson's article about me, in which he talked about what he was doing in the early 1950s. He took a group of students, and he did a lot of math, and then he thought it worked out, so he went to Fermi. Since Fermi was, at that time, considered to be both a theoretical physicist and a leading experimental physicist, he took some of his work with him to Chicago to explain it to Fermi. I'll send you a copy of his article about me. He said, after discussing it with Fermi, he knew he was going in the wrong direction. And then he said, "I learned more in an hour of discussion with Fermi than I learned in ten years with Oppenheimer." In fact, from then on, he didn't do that research again; he did statistical mechanics and other related stuff. So, if you ask me, this was his principled error, because he shouldn't have dropped it altogether. I don't think he should have stopped doing math either, especially considering he was already involved in it early on during his student years, as I heard.

Ji Lizhen: He was good at number theory.

Yang Chen-Ning: (He) had one or two articles published in Cambridge related to number theory.

Ji Lizhen: Very good.

Yang Chen-Ning: Do you know these articles?

Ji Lizhen: I know. I've heard about it. I kind of forgot. I've seen it before.

Politics Behind the Nobel Prize

Yang Chen-Ning: He became interested in physics later on, and of course, he made a big hit by proving the Feynman diagram, which was definitely Nobel-level work. I don't think it's fair that he wasn't given it.

Ji Lizhen: Yeah, there were four of them. The other three received it, but he didn't. There's a book called *QED and the Men Who Made It.*

Yang Chen-Ning: They can give two Nobel Prizes. If you ask me how I think about this, they gave it to Schwinger, Feynman and Tomonaga Shinichirō, and I don't think Tomonaga Shinichirō's work was anything comparable. I think part of the reason for the result is that the United States at the time used the atomic bomb to blow up Japan …

Ji Lizhen: A little guilty.

Yang Chen-Ning: A little guilty, so …

Ji Lizhen: Make up for it.

Yang Chen-Ning: And they gave Yukawa Hideki a Nobel Prize (1949).

Ji Lizhen: That's right.

Yang Chen-Ning: Then it was given to Tomonaga Shinichirō (1965). In terms of his contributions, he was nowhere near as good as Dyson.

Ji Lizhen: Yeah, and then I have this question. Is Dyson a little bit frustrated? He's just not very happy because he didn't get the Nobel Prize.

Yang Chen-Ning: Do you think he was unhappy that he didn't get the Nobel Prize?

Ji Lizhen: Yeah, a little bit. It's just that it's not fair.

Yang Chen-Ning: I don't think it's possible for him to be not unhappy at all. He's a very smart guy. He wrote a lot, and his tone in these papers is that he didn't care. But if you ask me, I think he did. That's just the way he was. Have you ever met him?

Ji Lizhen: I might have met him when I was at Princeton. I might have met him when I went to IAS for a year.

Yang Chen-Ning: When were you at the Institute for Advanced Study?

Ji Lizhen: I was there in 1994 and 1995.

Yang Chen-Ning: At that time, he was already …

Ji Lizhen: Retired.

Yang Chen-Ning: He was retired but never stopped working. Over the last few years, I think he wrote one book a year on average. He's very good at it.

Ji Lizhen: And he wrote a lot of reviews.

Yang Chen-Ning: Very good at reviewing books. Do you know Lin Kailiang?

Ji Lizhen: Yes, I know. Lin Kailiang wrote an article about him.

Yang Chen-Ning: Lin Kailiang wrote a very good biography of Dyson. Have you read it?

Ji Lizhen: Yeah, I read it a little bit. I haven't read it closely. I will go back and read it again.

Yang Chen-Ning: Because of that, he had some communication with Dyson via emails. Dyson had written an email, and Lin Kailiang published it. It seemed like he said that as for the three most important things in his life, one is … None of his three most important things is academic. In my opinion, he wasn't lying in his statement due to his own analysis, but I think he was deceiving himself because he can't be (not upset). He's a brilliant man. I knew that early on. Did you know that Cambridge University has an exam called Tripos (degree examination for undergraduates and some master students)?

Ji Lizhen: Yes, I know.

Yang Chen-Ning: If Dyson took the Tripos exam, I would never take it. There's no way I could compete with him.

Ji Lizhen: Senior wrangler, right?

Yang Chen-Ning: Yeah.

Reasons Behind Borel's Belief that Science Prizes Hurt

Ji Lizhen: I read what Dyson said about this in a book preface and felt he's not very happy either. He said that there were two kinds of mathematicians or scientists in otherfields. One kind solved a difficult problem and became famous, and the other kind who broke new ground and proposed new theories had great influence in the long run. He felt the latter was more important, which is my interpretation, because he didn't get a big prize, although his work was remarkable. Borel once came to the University of Michigan to give a lecture and said all the prizes should be eliminated. Why? Because they made a lot of people sad, and also because they made prize winners feel like they should get them. A lot of people got upset when they didn't get them. Then one of my colleagues said to me, was Borel kind of talking about himself? Was it because Borel hadn't won any big awards? So, taking these two things into consideration, I got a sense from Dyson that he was a bit unhappy. As you said, it wasn't really fair to him either.

Yang Chen-Ning: I think I agree with you, but I'm a little different from you. I think he has gone beyond the point of being happy or not. He is less emotional about himself. And there's something weird about him. The weird thing is … I don't know if you've seen Feynman's book.

Ji Lizhen: Which book? Is it *Surely You're Joking, Mr. Feynman!*?

Yang Chen-Ning: Yes, have you read it?

Ji Lizhen: I've read it twice. I like it very much. I think it's very interesting.

About Feynman and His Deeds

Yang Chen-Ning: I don't like Feynman.

Ji Lizhen: Because he gave the impression that he did everything so easily. I feel like sometimes he just appeared so smart, didn't he?

Yang Chen-Ning: He's not an easy person to be around. He's not a decent guy. He's very political.

Ji Lizhen: Oh, that's the way it was. Was this created on purpose?

Yang Chen-Ning: He was a contemporary of Schwinger, and they were both born in 1918, so there was a fierce rivalry between them when they were in their twenties. They were very different, and were both working on Feynman diagrams. Schwinger used traditional field theory while Feynman depended on guesswork. So they both later won the Nobel Prize at the same time. But they had totally different attitudes, and I wrote an article about Schwinger.

Ji Lizhen: Is there something about Feynman in the article?

Yang Chen-Ning: It's about Feynman.

Ji Lizhen: That's good.

Yang Chen-Ning: You can see that I don't like Feynman's style. He is contrary to the traditional Chinese way of being a man.

Ji Lizhen: Yes, he seemed to be very flashy.

Feynman's Unfriendliness to Dyson

Yang Chen-Ning: For him, there was no such thing as fairness. He didn't even have the concept. One of the things that I'm most unhappy about, and feel very strange is about Dyson. If you read all the books Dyson has written so far, he talked again and again about how much he admired Feynman, and he talked very strongly about this. If you ask Lin Kailiang, I think he would tell you in which article Dyson said that he worshipped Feynman as if he were a saint. But Feynman was very unkind to him. How do I know Feynman was very bad to him? It's because one year in Colombia there was a meeting. I sat at the dinner table and saw Isidor Isaac Rabi, who was 20 years older than me. In the 1950s and 1960s, in the field of physics in the United States, he had a close relationship with the United States government leaders. It is him who replaced Oppenheimer. Oppenheimer was the most famous and influential American physicist between 1945 and 1954, but in 1954—you know about the famous rumor—he lost his influence in the government, which had a big impact on him, so next came Rabi. I was talking about this at dinner that day back in the 1970s I said it was wrong that the Nobel Prize was not given to Dyson, but Rabi immediately said he didn't agree because he asked Feynman what contribution Dyson had made, and Feynman said Dyson had made no contribution.

Ji Lizhen: Really?

Yang Chen-Ning: At that time, I was shocked that Feynman came to this conclusion. First of all, this conclusion is completely opposite to the truth. Secondly, it's very Feynman.

Ji Lizhen: That's right.

Yang Chen-Ning: Feynman was an unreasonable man.

Ji Lizhen: Yeah, I could tell from reading that book that there was a chapter where Feynman was really dismissive of mathematicians. This was not good for Caltech's mathematics department.

Yang Chen-Ning: Yes, you were absolutely right, because he had the desire to dominate.

Ji Lizhen: Yeah.

Yang Chen-Ning: Someone wrote a biography of Feynman. There are several Feynman biographies. One book said very nice things about Feynman, but in the book, it mentioned that the author went to interview some of Feynman's colleagues at Caltech, and one of them was an experimental physicist. The experimental physicist said they were all a little afraid of Feynman.

Ji Lizhen: Yeah, Caltech's mathematics department couldn't advance; Feynman was responsible for that, I think.

Yang Chen-Ning: Because Feynman didn't think mathematics was important; he thought he could invent any mathematics he needed.

Ji Lizhen: Right. So, let's move on. We were talking about how to be creative.

Feynman's Most Important Contribution

Yang Chen-Ning: I would say Feynman was creative, and that's true, but his greatest contribution was not actually that. His greatest contribution was the path integral.

Ji Lizhen: Not the Feynman diagram?

Yang Chen-Ning: Path integral, I think, captures the true spirit of quantum mechanics. Up to now, mathematics people haven't figured it out yet.

Ji Lizhen: Now it's very mathematically difficult for us because when you calculate an integral in an infinite space, how do you reduce it to a finite dimension? How do you do that? That's a big question.

Yang Chen-Ning: You know, before World War II, in the 1930s, there was the delta function in Dirac's book.

Ji Lizhen: Right.

Yang Chen-Ning: This delta function was not acceptable in mathematics at that time. But during and after the war, Laurent Schwartz turned it into distribution theory. He won the Fields Medal for his work. But turning the delta function into distribution is not as difficult as the path integral. Path integrals involve an i problem. If there is an i at the point of the action, it becomes a phase, and without that i, it's necessary to do some study on probability theory. However, with and without i, this is an insurmountable difficulty. I think there is something very profound going on here.

Ji Lizhen: Yes, it's easier to converge without i.

Yang Chen-Ning: Without i, mathematics would be very easy.

Special Relativity: Einstein, Lorentz and Poincaré

Ji Lizhen: That's right. Let's move on. I have another question here, about originality. Everyone agrees that Einstein was creative, but the question is what made Einstein so creative?

Yang Chen-Ning: In many aspects, he didn't just create one idea.

Ji Lizhen: Yes, can you tell us what the factors are, and why Einstein's ideas are so original?

Yang Chen-Ning: The first major originality is… Einstein did not invent the term *relativity,* which I wrote in an article.

Ji Lizhen: Was it Lorentz, or who?

Yang Chen-Ning: Poincaré.

Ji Lizhen: Yes, Poincaré.

Yang Chen-Ning: Regarding the Lorentz transform in special relativity, Einstein didn't propose it; Lorentz did. So, all of these things were already there, and I wrote an article which I'll email you later. I mentioned that Lorentz did the math, but he didn't understand the mathematics. He changed it from T to T', thinking that T' was a—he gave it a name—virtual time or something. In short, what Einstein did was T', not virtual time, but another person's T. That is, a different T for each person.

Ji Lizhen: The theory of relativity came out because people are different.

Yang Chen-Ning: So, in other words, Einstein introduced this idea into Lorentz's mathematics and said what he meant, which Lorentz didn't understand. Later, Lorentz said in his later years that this was where he failed. My article on Einstein quotes from Poincaré's 1904 article. Poincaré said relativity meant that different observers perceived things differenty. That is, the length of a thing one measures was not necessarily equal to the measurement by another person, but they came up with the same law, and this is called relativity. The statement seems absolutely true to you now, but he did not understand its meaning.

Ji Lizhen: The physical sense.

Yang Chen-Ning: He was correct in the philosophical sense. He spoke of the philosophy of relativity, but he couldn't put it into practice. Einstein put the two together, so my article said, "Poincaré got the philosophy, Lorentz got the mathematics, but Einstein got the physics."

Ji Lizhen: Yes, he grasped the essential things.

Yang Chen-Ning: People were amazed after they understood it. Well, I think it's because Poincaré and Lorentz were the most important people at the time, but Einstein was a nobody, so that's very impressive. And then his second big (original) work was perhaps even more important. According to Einstein himself in his later years, this was his only revolution. In 1905, he said that electromagnetism was a particle and light was a particle. At the time people must have thought that was nonsense. Why? Because interference was there. Interference was done in the 19th century, and suddenly you were saying it's a particle, which is simply unorthodix. So, the other thing that's really interesting, you know, is that by 1913, because of the general theory of relativity, everyone knew this guy was a great young man, so Max K. E. L. Planck wanted to get him from Zurich to Berlin, and he wrote to the dean of Berlin, asking the school to invite Einstein to come. He wrote that this young man had done first-rate work in all sorts of fields, but suddenly he turned around and said that of course he had sometimes done absurd things, for example, about light, but Planck went on saying that we should bear with him even if he was a little wrong, and this was light. Einstein, however, was unimpressed. In 1916 and 1917, he wrote two more papers to make the particle view of the photon more concrete. That work was the originator of laser which induced radiation emerges.

Ji Lizhen: Really?

Yang Chen-Ning: So, he understood light sharply. And then the third (big original concept) emreged in 1924. There was a young Indian named Satyendra Nath Bose, and today Bose is famous because of Bose–Einstein.

Karen Uhlenbeck's Father-in-Law

Ji Lizhen: Bose–Einstein, yeah, I've heard of it, but I don't know its story.

Yang Chen-Ning: How did this Bose–Einstein come into being? Bose wrote a letter to Einstein. He said, "I wrote an article and sent it to England. They refused to publish it. Could you please translate it into German and publish it in Germany?" This guy was a little brusque, but Einstein actually translated it into German and published it in a German journal, under Bose's name, but he put a footnote at the bottom, saying this was a very interesting article, and he wanted to say a little bit more about this topic. And then two or three articles came out in a row using the idea of Bose, because Bose was talking about light. I think Bose was very insightful, and he knew that Einstein would be interested in light. He used another way to study light, which was to talk about the statistics of light. He kind of figured out that Einstein would be interested in this article, and he was right; he had a lot of insight. Turns out Einstein looked at it and said he had a good idea, but why not just use light instead of doing the same thing with particles of matter? So, he said that if a particle of matter was calculated using the Bose method, it would condense. What is condensation? It means that particles of that kind can change into liquids without interaction and can also change from gas to liquid. After this article came out, all physicists thought that Einstein was confused—how could there be no interaction? At that time, it was thought that liquids were attractive, and how could they become liquid without attraction? One of the scientists who refused to accept this idea is one of his best friends, Paul Ehrenfest, who was in Leiden and later committed suicide, was a Viennese Jew who had lived in Leiden for a long time and was a good friend of Einstein. He taught one of his students, George Uhlenbeck, who was Karen Uhlenbeck's father-in-law.

Ji Lizhen: Well, then he came to Michigan.

Yang Chen-Ning: Where is Karen Uhlenbeck now?

Ji Lizhen: She's retired now.

Yang Chen-Ning: Where was she before her retirement?

Ji Lizhen: She was in Texas.

Yang Chen-Ning: Karen Uhlenbeck is George Uhlenbeck's daughter-in-law, a graduate student at Leiden University, who was assigned by his thesis adviser to write an article pointing out that Einstein was wrong. George Uhlenbeck was later considered by many to have deserved a Nobel Prize, because it was him who first discovered spin. But by the time I was a graduate student, by the time I did my dissertation, in the 1950s and the 1960s, it was clear that Einstein wasn't wrong. His proposal of Bose–Einstein condensation turning into something liquid without interaction was considered impossible because the condensation required a very low temperature. However, this was finally achieved in the 1990s, and by then Bose–Einstein condensation had become a reality, winning several Nobel Prizes. So, Einstein had a knack for getting to the bottom of something that no one else had thought of.

Ji Lizhen: Yeah, now the question is, because Einstein was so innovative, can we learn something from his work and his way of doing things?

Einstein and General Relativity

Yang Chen-Ning: I haven't talked about general relativity yet.

Ji Lizhen: Yeah, that happened later.

Yang Chen-Ning: How did general relativity come about? Because when special relativity came out, there was no idea of symmetry. As a result, the famous number theorist in Zurich …

Ji Lizhen: In which area? Hermann Minkowski?

Yang Chen-Ning: Minkowski wrote an article saying that Einstein's special theory of relativity had symmetry, which is namely, SO(3, 1). Right after Einstein saw that, he wrote an article saying he didn't like it, and he gave Minkowski's idea of symmetry a name: unnecessarily subtle.

Ji Lizhen: Really?

Yang Chen-Ning: But a year or two later, he changed his mind. He no longer thought it was unnecessary. When he looked back at that time in his later years, he said that it took him a year or two to understand the importance of symmetry. Later, he thought that he wanted to make it bigger, so he needed to find a formula. After writing out the formula, he could use any transformation to produce a new formula, which would be the same as the original formula. He specialized in this from 1907 to 1914, with a little help from Marcel Grossmann, and finally, he came up with the theory of general relativity. And that, of course, was the other thing that made him great.

Ji Lizhen: Yeah, the thing that made him famous. Now I wonder, what's the thing that a lot of people are interested in? You mentioned it a little bit. Can you now summarize what made Einstein so innovative and able to see things that other people couldn't see? What can people learn from him if they want to be a little bit more creative?

Yang Chen-Ning: I think it's like great poets, like Li Bai and Du Fu. How can they …

Ji Lizhen: Talent, it has something to do with talent.

Yang Chen-Ning: I think human beings' brains are marvelous.

Ji Lizhen: It's just natural.

Yang Chen-Ning: Yes, of course. The strange thing is that God didn't create one brain but countless brains, and most of them don't have this characteristic.

Ji Lizhen: The following questions may have more to do with mathematics, because mathematics is more important. For the average student, there are two types of books to read. One is textbooks, the other is extracurricular books. What do you think of the importance of these two kinds of books, and the difference between them?

Yang Chen-Ning: In my opinion, if your goal is to cultivate, say, a first-rate mathematician, then I think …

Ji Lizhen: Does none of that matter?

How to Train Top Students

Yang Chen-Ning: How can we train first-class people? I think, firstly, a necessary condition is that he has to be gifted. Gifted in what ways? Of course, this is another question. So, the question actually is: how do you best raise a gifted child? In other words, being talented is a must. As for a gifted child, I think it is best to help him cultivate his interests at an early age. For example, Dyson, he wrote so many books. After you read them, you admire them, but I think …

Ji Lizhen: He wasn't so influential.

Yang Chen-Ning: There's no permanent value in human history. I think he could absolutely do first-class work and obtain immortal fame. You know, I like referring to Du Fu's "writing is a deed of eternity," which means that good writings can last one thousand years. I think Dyson's work could have remained through the ages of mathematical work, but he didn't make it.

Ji Lizhen: So, he got his focus wrong.

Yang Chen-Ning: Because of his great abilities and wide-ranging interest, he went in other directions.

Ji Lizhen: That's right. You may know Maxim Kontsevich who also won a Shaw Prize. Three years ago in Sanya, I organized a student conference and invited him to come. Later, when we got together, he told me that his teacher was Gelfand, and Gelfand expected him and other fellow students to carry around the complete works of Euler and things like that.

Yang Chen-Ning: Leonhard Euler?

Ji Lizhen: Yeah, they all had to carry the complete works of Euler on our backs. Gelfand had said that he didn't want them to read the complete works of Euler, but to carry Euler's book, and Euler's spirit would permeate them. What do you think of this? It sounds a bit mysterious to me.

Yang Chen-Ning: I think his purpose was to make a metaphysical statement, and it makes sense to me.

Ji Lizhen: It makes sense? Did he want his students to learn from Euler or something else?

Yang Chen-Ning: I think this is closely related to people's mindsets.

There was once a direction of exploration that Euler was best at, and Gelfand thought that students must learn it. It's not something that can be mastered after one proved Euler's work one by one. It's somehow you have got to get the spirit.

I think that's what Gelfand meant. Gelfand had no problem in mathematics; he was absolutely first-rate.

Ji Lizhen: Yes, his status is very high.

Yang Chen-Ning: But what did he do that would be a deed of eternity like writing?

Ji Lizhen: Yeah, none of them were simple. Gelfand studied a very broad range of things. I don't know him very well. He studied many things, like the Cusp form, an automorphic form. You said Schwartz did distribution theory, so did Gelfand. He wrote six books, one of which was *Generalized Functions*.

Yang Chen-Ning: All he did was analysis (field)?

Ji Lizhen: It's all related to analysis. It's related to group representation.

Yang Chen-Ning: Did he also do topology?

Ji Lizhen: Take topology, for example. The index theorem was proved by Atiyah and Singer, but it was Gelfand's guess that first raised the question; its index should be a topological invariant and he posited how to find it.

Yang Chen-Ning: Do you think Gelfand already had the idea of an index?

Ji Lizhen: Yes, I do.

Yang Chen-Ning: Already?

Ji Lizhen: Because in functional analysis, there is a Fredholm operator that calculates the difference between the dimension of the kernel and that of the co-kernel. Because he knew that the index of this Fredholm operator didn't change with a little perturbation, he already knew that there should be such a theorem, so Atiyah and Singer proved it. And in a sense, Gelfand contributed because he made it clear that this thing should be a topologically invariant. Alexander Beilinson, who was now his student, told me last time that a lot of Gelfand's work was collaborative because he had ideas that he wanted people to pursue. If you want me to give a specific example, I don't know. Anyway, he has done a lot in many fields, and he can open people's eyes to see something. Therefore, his influence should be very big.

Yang Chen-Ning: That's right. I have a question for you, in the overall history, who do you think was the more influential? Gelfand or von Neumann?

Ji Lizhen: Mathematically?

Yang Chen-Ning: Yes.

Ji Lizhen: I think it's Gelfand.

How to Compare Gelfand and John von Neumann

Yang Chen-Ning: Do you think Gelfand has greater influence than von Neumann?

Ji Lizhen: Yeah, because von Neumann is best known mathematically for von Neumann algebra, which I think is a little bit limited mathematically. Some people who do von Neumann algebra, of course, think it's important. But if you ask other mathematicians, for example, they think Alain Connes has now moved on to noncommutative geometry.

Yang Chen-Ning: He discovered game theory.

Ji Lizhen: That's right. von Neumann of course was a big influence, and I think von Neumann's influence was very broad. That's why it's so important to evaluate these things, which leads to my next question. I have prepared a total of 39 questions, and now I have just asked the third one.

Yang Chen-Ning: Did I ever tell you how I met von Neumann? Because von Neumann liked big parties, and big parties had a lot of drinks. Even though I didn't drink a lot, I would go to his house. He liked to tell stories, and a lot of them were dirty jokes. He liked to tell dirty jokes and then laugh. He was, of course, a very clever man, and because he had a very good memory, he was also a very interesting man. I went to the Institute for Advanced Study in 1949, and I thought by then he was being polite to other mathematicians at the institute because he was working on hydrogen bombs.

Ji Lizhen: Was he working on the supercomputer too?

Yang Chen-Ning and the Unit Circle Theorem

Yang Chen-Ning: These two things were related because his computer work was related to the hydrogen bomb, and the work was mainly conducted by him. I don't think he did mathematics work anymore. I really think so, and I don't think he had a lot of discussions with Dean Montgomery or Selberg. But I went to see him because I had stumbled upon a theorem with Lee Tsung-Dao, and it's a famous theorem now, even though it's only a small theorem.

Ji Lizhen: What kind of theorem?

Yang Chen-Ning: It's a high school algebra theorem. It's now called the unit circle theorem. Do you know it?

Ji Lizhen: I don't know.

Yang Chen-Ning: Do you know a very famous man in a research institute in Paris, called David Ruelle?

Ji Lizhen: Yes, I know his name.

Yang Chen-Ning: Do you know David Ruelle?

Ji Lizhen: He wrote a mathematical proof, thermodynamics, and zeta function.

Yang Chen-Ning: Complexity and things like that. He's famous for that. He wrote a little book.

Ji Lizhen: *The Mathematician's Brain.*

Yang Chen-Ning: Yeah, there is one chapter talking about the unit circle theorem in the book. The unit circle theorem came into being when Lee Tsung-Dao and I worked on statistical mechanics. Suppose there are n spins and they interact with each other, then their property is determined by a polynomial. The polynomial can do n spins. When they interact with each other, there's a polynomial, and that's what we were working on. By studying the first two spins, we found that, if the interaction is attractive, the polynomial is a quadratic polynomial whose roots lie on the unit circle. If it is not attractive, it may not be on the unidimensional circle. Then it becomes three rotations, and if all interactions are attractive to each other, all the roots are on the unit circle. We said this is good. So if we make four and five spins, then, of course, the more the spins, the more roots there are because the order of a polynomial is equal to the number of particles. After we had done four or five, we came to the conjecture that, as long as interaction is attractive, no matter how many roots there are, they are all on the unit circle. This is now known as the unit circle theorem. But we did not know how to prove it, we just made some possible trials, and then we thought about it for several weeks, and then we went to somebody—I think it was von Neumann—and then von Neumann said, well, polynomials have roots on the unit circle, and that's a popular topic in number theory. He said, "I'll refer you to a book you should read. It's Hardy's *Inequalities*."

Ji Lizhen: That's right. It was written by Hardy, Littlewood and George Pólya.

Yang Chen-Ning: Yes, there is such a book.

Ji Lizhen: Yes, it's very famous. It was written by three people.

Yang Chen-Ning: We read that book, and it would tell you the types of n-order polynomials. My impression was that it is a sufficient condition that its roots were all on the unit circle. But the bad thing was that the sufficient condition for polynomials of order n, which said that all the roots were on the unit circle. There are n inequalities. Well, we could prove this inequality for $n = 2$, and prove these three inequalities for $n = 3$, but we couldn't do that for $n = n$, so it's completely useless. That's the only time I approached (a mathematician). I only consulted mathematicians once or twice in my 17 years at the Institute for Advanced Study, and this was one of them with no result. The other time it worked, and I wrote about this later in an article; it was Hassler Whitney. This was in the 1960s, and I was working on something. I guessed there was a certain kind of equation with roots, so I went to Whitney. He told me there was an index theorem, so I checked the index theorem. Sure enough, the equation I wrote down was consistent with the index theorem, so I had a root. This was a successful consultation. No other consultations were made, except in the 1960s, when I went to Borel. Did I tell you that story?

Ji Lizhen: Yeah, it's about Lie groups.

Yang Chen-Ning: Yes, as a result, he taught us physicists the fixed-point theorem. It was very beautiful. The simplest fixed-point theorem is from the early 20th century, isn't it?

Ji Lizhen: That's Luitzen E. J. Brouwer. Isn't it Brouwer?

Yang Chen-Ning: Yeah. It's from the 20th century?

Ji Lizhen: Yeah, the early 1900s.

Yang Chen-Ning: It was after Poincaré.

Ji Lizhen: Yeah, Brouwer was a little bit later than Poincaré.

Yang Chen-Ning: The topology that Poincaré studied was quite complicated.

Ji Lizhen: No, his topology was also algebraic, and so was Brouwer's. Brouwer's story is interesting too, you know. He was interested in this kind of mathematical basis, and he had to prove that he was capable, so he would do a little mathematics to show you, and he did a couple of famous things on topology, then he went back. That von Neumann you were talking about, well, von Neumann turned out to be pretty miserable in mathematics. You know, in 1900, Hilbert asked 23 big questions, and then in 1950, at an international congress of mathematicians, they invited von Neumann, who was the biggest mathematician at the time. They asked him to propose some questions about mathematics, and to explain what the world of mathematics was like. Expectations were so high that all the rooms were full, but when von Neumann arrived, he didn't seem prepared. He didn't talk about any questions at all. He just kept talking about his old work. After he finished talking, there was not much reaction in the room. Von Neumann lowered his head and left. Everyone was disappointed, and so was he.

Yang Chen-Ning: Really, I didn't know that story.

Ji Lizhen: Yeah, I heard that.

Yang Chen-Ning: I think you were talking about a gathering in 1950, weren't you?

Ji Lizhen: Yeah, he was already …

Yang Chen-Ning: I think he was already working on the hydrogen bomb.

Ulam and John von Neumann

Ji Lizhen: Yeah, he wasn't doing mathematics anymore by then, but his reputation was there, so expectations were still high. I remember that impression, and I read a biography by Ulam about von Neumann. As you said, he liked to party, and then his wife cheated on him with one of his assistants, which made him very sad. It was like he was partying all the time to please his wife. One of his assistants came often, and took away his wife.

Yang Chen-Ning: What about Ulam's relationship with von Neumann?

Ji Lizhen: It's complicated.

Yang Chen-Ning: There was a mathematician at Stony Brook University who was a dozen years younger than me, Ronald G. Douglas. Do you know this guy?

Ji Lizhen: I know him.

Yang Chen-Ning: He went to Texas.

Ji Lizhen: That's right.

Yang Chen-Ning: I don't know if he's still there.

Ji Lizhen: Yes, still there.

Yang Chen-Ning: He was very unhappy with Ulam. Ulam wrote articles and books about von Neumann. I remember Douglas was very unhappy. Douglas was very impressed with von Neumann, and he felt that Ulam was making absolutely no sense. Did I tell you that Ulam was a very strange man himself?

Ji Lizhen: Yeah, I've read some of his articles.

Yang Chen-Ning: Gian-Carlo Rota wrote an article about Ulam.

Ji Lizhen: Yeah, I've read it before. Ulam also has an autobiography, and I can understand why Douglas admires von Neumann so much. It is because he did C* algebra, von Neumann algebra, so people in the field think highly of von Neumann, but it's hard to say what other people outside think of him.

Yang Chen-Ning: So, I think it's fair to conclude that the general mathematician doesn't think von Neumann's mathematical talent has fully emerged.

Ji Lizhen: Yes, because he left mathematics too early and he died too young. I think he's great. I also heard a story about him. You know, when Pólya was teaching at ETH Zurich, he used to check every day before class to see if this student, von Neumann, was there, and he had a fear if von Neumann was there because this student was so good that he would really embarrass the instructor in class. Pólya was also a very clever man.

Yang Chen-Ning and the Pre-fame Raoul Bott

Yang Chen-Ning: There really is something strange about being a mathematician. Let me tell you two stories. One happened when I was at the Institute for Advanced Study. I'd been there for 17 years, and Oppenheimer came there in 1947. When he just arrived, his youngest daughter was still in kindergarten, and his wife organized a small kindergarten for children whose parents were from the institute. Later, all three of my children went to that kindergarten, so we got to know the teachers of the kindergarten very well. One teacher of the kindergarten said one day that her kindergarten was very small (I think usually there were only 20 kindergarteners), and she said, "We have children with anxiety in this school all the time," and "I know all the children with anxiety are kids whose fathers are mathematicians." And I think that makes sense, because mathematicians are kind of weird, and because mathematicians are different from physicists, they often split hairs. And when they do so, the way they deal with their kids is sometimes, well .... So anyway, it seems that mathematicians' children often don't fit in easily. I'll tell you another story. Bott was about my age, and he spent a few years at the Institute for Advanced Study before he went to Harvard. I had become a professor at the Institute for Advanced Study in 1955, and my eldest child was in the kindergarten with his children. The kindergarten had the parents help repair things on Saturdays and Sundays. For example, there was a broken shelf and needed to be fixed, so I went to fix it one Sunday, and Bott was also there to fix it, and that's how we met. We had known each other but hadn't talked much before then. That time we did. I remember it very well because he hadn't gotten tenured yet and hadn't gone to Harvard, so he was probably a little jealous. He said things like, "How do you feel about getting tenured?" At that time, I didn't think he had done his most important job, which was probably not done until he went to Harvard.

Ji Lizhen: No, that was done at the University of Michigan.

Yang Chen-Ning: He went to the University of Michigan first and then to Harvard?

Ji Lizhen: Yeah, because when he was at Harvard, he did the Bott periodicity, and he became famous. I also heard a story that after he did this job …

Yang Chen-Ning: He was in electrical engineering; didn't he do it at Carnegie Mellon? From there, he went to the Institute for Advanced Study, where he transferred to mathematics, and then he went to Michigan.

Ji Lizhen: Yeah. Well, Weyl went to Carnegie Mellon to give a colloquium talk. Bott's tutor invited Weyl to give it. And then Bott went to the colloquium dinner. Bott and his adviser made a theorem on electronic circuits, probably graph theory, which they shared with

Weyl. When Weyl heard about this, he thought it was good and encouraged Bott to go to the Institute for Advanced Study. Then he went to Michigan. After his good work was done, Harvard offered him a position. Bott asked his father what he thought of the offer, and his father said, "In a family like ours, you can't turn down a job offer from Harvard." You were talking about the difference between a mathematician and a physicist, and that's one of my questions because I've had that experience. I went to IAS (Institute for Advanced Study at Princeton) for a year in August 1994, and when I arrived on my first day …

Yang Chen-Ning: Did you say 1994?

Ji Lizhen: 1994, yes.

Yang Chen-Ning: You already had a Ph.D. by then.

Difference Between Mathematicians and Physicists

Ji Lizhen: Yes, I was visiting as a Ph.D. I was on a short-term basis, just for a year. We went to the Fuld Hall at the Institute for Advanced Study, and there was a security guard to pick up his keys or something. The security guard was kind of weird. When he met me, he said, "Don't tell me what college you're from yet. I can guess." When I asked him what the difference was between a mathematician and other disciplines, he said mathematicians were a little bit strange, just a little bit crazy. He said that physicists were more normal, and then he said that historians were more like scholars. You think he's basically right, don't you? When I came back and told my parents about it, my dad agreed and said that we mathematicians were kind of weird.

Conventional Chinese Mathematicians

Yang Chen-Ning: Regarding that, I don't think Chinese mathematicians are very eccentric compared to the others.

Ji Lizhen: Really?

Yang Chen-Ning: For example, very few American mathematicians ever become university presidents, but many Chinese mathematicians do become university presidents. In fact, if you ask, since the founding of China, what percentage of university presidents have studied mathematics, physics, biology, chemistry, I would suspect mathematics is actually the most popular.

Ji Lizhen: What's the reason for that?

Yang Chen-Ning: Why is that? First of all, the idea of mathematics in China is derived from the public perception and students' own opinions, which are different from those in the West. In China, some university presidents, mathematician university presidents, learned in a disciplined and orderly manner. Many eccentric scholars in the United States are mathematicians, who will never become university presidents because they do not want to work in those positions. Society really does not want them to be presidents either.

Ji Lizhen: Yeah, like the ones who have been making a big fuss, like Grigori Perelman. Look what a strange mathematician he is. And Alexander Grothendieck, too. You've heard of him?

Yang Chen-Ning: Yes.

Ji Lizhen: Yes, you see, he is also a very intelligent person, but he is very strange.

Yang Chen-Ning: I think it has something to do with the whole Chinese custom, the Chinese cultural tradition, because the Confucian cultural tradition discourages strange behavior, so it probably has a lot to do with that.

Ji Lizhen: Sometimes originality and weirdness go hand in hand. A quick question I'd like to ask you is whether there are any weird people in physics?

Yang Chen-Ning: Physicists?

Ji Lizhen: Right.

Yang Chen-Ning: Of course.

Ji Lizhen: Can you give me an example?

Yang Chen-Ning: Feynman.

Ji Lizhen: Feynman? Compared to Perelman, Feynman was smooth, resourceful and good at whitewashing himself. But Perelman was not.

Yang Chen-Ning: When you say strange, do you mean that Chen Jingrun is strange or Grothendieck? Which one are you talking about?

Ji Lizhen: Speaking of strange, I'll give you an example, like Perelman.

Yang Chen-Ning: Who?

Ji Lizhen: Perelman, the man who solved Poincaré conjecture.

Yang Chen-Ning: Right.

Ji Lizhen: I think he's totally different from Feynman, who presented himself in a very good light, but Perelman wouldn't do that, and I don't think Grothendieck did either.

Yang Chen-Ning: There's no one like Chen Jingrun in physics, nor Grothendieck in physics.

Ji Lizhen: My feeling is that people in physics seem to know a lot about the world, especially about emotions, and some people in mathematics are a little bit "on a different planet," aren't they?

Yang Chen-Ning: That's right.

Ji Lizhen: Okay, let's move on to the third question, the third question of our 39 questions. We were talking about students. Is that relevant? In Chinese culture or Western culture, what is a good student, and what is a bad student? You are in contact with a lot of people, so I don't think this definition is only based on the students' academic performance, which is also very important to be a genuinely good student. That is, how can a student become a truly good student during studying? What do you think of this question?

Chinese and Western Values

Yang Chen-Ning: I think there's a big difference. The values in the Chinese cultural tradition are quite different from those in the West. I can also recommend you an article on this to read. Chinese are influenced by the Confucian traditions, and it is necessary to set up an image, the highest level of which is the gentleman (君子). Today, I think, when discussing a young person's achievements in Chinese mainland, Chinese Taiwan or Hong Kong SAR, this person's attitude toward being a good person is much more important than that in the West. Westerners consider this aspect relatively less important. I wrote an article about Fermi, and I will send it to you. It is about how much I admire Fermi, not only for his knowledge but also for his attitude toward integrity.

Ji Lizhen: Yeah, I remember he did calculations to help prove a theorem. Yes, I thought he was very friendly.

Yang Chen-Ning: Yes, there is a story in my article which gave me a deep impression. It was back in the 1950s, when Oppenheimer was very important as an adviser to a very powerful government agency in the United States called the General Advisory Committee (GAC). The GAC was the institute that advised the United States government on all things concerning nuclear energy and nuclear weapons. So at that time, in the late 1940s and the 1950s, it was the most powerful institute. I remember very well that Oppenheimer, who was the chairman of this committee, went to see Fermi one day and then came back to me, because he knew I was very close to Fermi. He said he had just gone to see Fermi because his membership in the GAC had expired, and he had tried to persuade him to continue, but Fermi refused. After he talked to Fermi for a long time, the latter finally said, "you know, I don't always have confidence in my own viewpoint on these things, so I'm unwilling to insist." Oppenheimer told me the story, and I wrote about it in this article. This spirit of Fermi (adhering to principle) is the highest principle of being a man in the Chinese tradition. When such an important committee would ask someone to continue, the average American would be eager to do so. Therefore, in this regard, I would say Fermi is like a Confucian scholar, but there is no such perspective in the American tradition, so there is a question to consider: Is it detrimental to breakthroughs in academic research if too much attention is paid to Confucian demeanors? I think this is a very important question. In my opinion, it may be that people are too constrained by wanting to be decent people. The whole person from the very beginning is constrained, with too much self-reflection. The United States does not pay as much attention to this sort of self-monitoring.

Ji Lizhen: I don't think Einstein was that kind of person, was he? What do you think?

Einstein's Personality

Yang Chen-Ning: As far as I know, Einstein's personality, except for his relationship with women, was not necessarily bad. At least I don't think it was bad, but I don't think Einstein was an ambitious person.

Ji Lizhen: Why do I bring this up? Because I recently read a short biography of Einstein written by Snow.

Yang Chen-Ning: Who wrote it?

Ji Lizhen: Snow, the one who also wrote the preface to Hardy's book.

Yang Chen-Ning: When was this? Is it new?

Ji Lizhen: It's old. In the 1960s, Snow.

Yang Chen-Ning: The 1960s?

Ji Lizhen: Right.

Yang Chen-Ning: Charles Percy Snow, you say?

Ji Lizhen: Yes, Charles Percy Snow's biography of Einstein.

Yang Chen-Ning: Yeah, forget about that one. It's not very well written.

Ji Lizhen: It's not good, is it? I was surprised by his portrayal of Einstein.

Yang Chen-Ning: Because he didn't understand Einstein at all.

Ji Lizhen: Really?

Yang Chen-Ning: Snow wrote very well about Hardy.

Ji Lizhen: Right. Did you say Einstein's (biography) was not well written?

Yang Chen-Ning: I think you said that there's a novel, called *Variety of Men*, written by Charles Percy Snow, and the worst one is about Einstein.

Ji Lizhen: Because what he wrote is surprising. I read a passage …

Yang Chen-Ning: Because he didn't understand Einstein. I don't think he understood Einstein's physics, and he didn't understand Einstein's life, either. But Charles Percy Snow does write very well about Hardy.

Ji Lizhen: Yeah, we all knew Hardy because he was famous, and then we read this book. You were talking about Einstein not being very good with women. It's also in the book, in his biography, that Einstein's first wife was not very beautiful. Why did he marry her? You can take a look; I read it and was also very surprised.

Yang Chen-Ning: Actually, we all know the reason.

Ji Lizhen: Really? What's the reason?

Yang Chen-Ning: Because they were classmates. When he was courting his wife, some of his letters showed what he was thinking at that time. He thought that when he married his wife, he could talk about physics all the time, so he took a fancy to his wife because they probably had a good time talking about physics together.

Ji Lizhen: That's not what Snow wrote at all.

Yang Chen-Ning: I don't remember. I just remember the book that I read, and I don't remember who else was in it. I just remember he wrote very well about Hardy and very badly about Einstein.

Ji Lizhen: Yeah, and he wrote about Stalin. What he wrote was surprising and no one said that about Einstein. He wrote that Einstein was a strong man with very strong desires and so on. I was also very surprised to read that Einstein looked very muscular. Well, what you just said about how to be a good student and a good person is very interesting. It's about six o'clock. When are you going to leave?

Yang Chen-Ning: Yes, I do need to leave now. We'll chat next time.

The Third Interview

Time: March 6, 2017
Place: Institute for Advanced Study at Tsinghua University
Interviewer: Ji Lizhen, Wang Liping
Recorder: Wang Liping
Collator: Peng Cheng, Wang Liping, Ji Lizhen

Before the interview, we discussed with Dr. Yang about the mathematical problems of modular functions and modular groups, which are among the most beautiful parts of mathematics. In the 19th century, every competent mathematician wanted to devote himself to the study of elliptic and modular functions. Gauss, for example, was the first mathematician to explore in this area, and his famous work inspired the work of Abel and Jacobi on elliptic functions and Abelian integrals for which Abel was famous. After conducting relevant research, Weierstrass went from a high school teacher to a professor at the University of Berlin, and the most famous work of Riemann was related to these problems, which frustrated Weierstrass for many years. The academic battles between Abel and Jacobi, and between Riemann and Weierstrass, are fruitful examples and, of course. Poincaré and Klein's famous academic battle was related to automorphic functions, which were generalizations of elliptic functions. The Wiles solution to Fermat's Last Theorem also depended on it. No wonder Dr. Yang wanted to take this subject even at his age. But this conversation about math was just the beginning. Next, we discussed why Wu Chien-Shiung didn't win the Nobel Prize, and how scholars at the Institute for Advanced Study at Princeton, the purest ivory tower of all, and especially André Weil, competed with other mathematicians. Moreover, we discussed the academic battles between Hua Loo-Keng and Siegel, and between Yang Chen-Ning and Feynman, the most important physicists and mathematicians of the time. This interview also discussed love stories between Nobel and several women, the meaning of life, and the relationship between religion and science.

Discussion on Elliptic and Modular Functions

Yang Chen-Ning: In a small rectangle, there will be some repeated functions called modular function.

Ji Lizhen: No, it's called an elliptic function.

Yang Chen-Ning: An elliptic function? Is it elliptic or modular?

Ji Lizhen: One more step, and you can have a lot of discrete groups, all of which form a space, a modular space.

Yang Chen-Ning: Is it an Abelian group?

Ji Lizhen: Yeah, if you think of all the Abelian groups put together, everything put together to form a modular space, then the functions on the modular space are called modular functions.

Yang Chen-Ning: Okay, and then the next one is the upper half plane, then you use a discrete finite group partition, and it becomes a block, albeit a weird one …

Ji Lizhen: Hyperbolic.

Yang Chen-Ning: Weyl appreciates this very much. The modular function inside, which is the transformation under the group, is invariant and becomes the same thing in pieces. Is this development due to generalized elliptic functions?

Ji Lizhen: That sounds right. Can you explain that a little bit more?

Yang Chen-Ning: In other words, there is a transformation, z' is equal to $(az + b) / (cz + d)$. Why do you want to study this transformation?

Ji Lizhen: It's like this: because when you're at this point, SL(2, Z), let's say you're given one …

Yang Chen-Ning: SL(2, Z) is easy to understand, but why would you write this transformation?

Ji Lizhen: Well, how do you look at the translation group? There are two parts or periods, one is $\omega_1 z$, and the other is added $\omega_2 z$ to it to get the lattice $\omega_1 z + \omega_2 z$. You can take ω_2 out, then we get $(z + \omega_1 / \omega_2 z)$, where $z = \omega_1 / \omega_2$ is the point on the upper half plane. We can take the periods $a\omega_1 + b\omega_2$ and $c\omega_1 + d\omega_2$ of the lattice and get a different point in the upper half plane $(a\omega_1 + b\omega_2) / (c\omega_1 + d\omega_2) = (az + b) / (cz + d)$. That is why the different points in the upper half plane are meta-related to the modular group, and their corresponding Riemann surfaces are not different.

Yang Chen-Ning: I see.

Works of Langlands and Other Mathematicians

Ji Lizhen: Because I thought Langlands was a very thoughtful mathematician, I wrote to him and said I'd like to choose some of his review papers and translate them into Chinese. He said fine, and I asked him to choose, but he declined and said, that he would let me do it myself, and when I was done, I would show him. After I had made my choices, I went to consult some of his more important disciples and asked them to do a second-round selection. Later, four or five of us discussed our choices and I showed Langlands the selected pieces. He said okay. Then he wrote a 30-page preface.

Yang Chen-Ning: So in English? Is the whole book in English?

Ji Lizhen: It needs to be translated into Chinese.

Yang Chen-Ning: Is it in both English and Chinese, or just Chinese?

Ji Lizhen: The English one is the preface that he wrote specifically for our book, and we had it translated into Chinese. Both the Chinese and English prefaces are in the book, but they haven't been published yet because of some delays.

Yang Chen-Ning: Please let me know when the book comes out and I'll buy one.

Wang Liping: When it comes out, I'll give you one copy.

Yang Chen-Ning: Did you publish *The Biography of Hua Loo-Keng* written by Wang Yuan?

Wang Liping: It was not published by us. It was published by the Science Press.

Yang Chen-Ning: Chern Shiing-Shen has many books published in English. Have you ever published a selected copy about his research work?

Ji Lizhen: The World Scientific Publishing Company has published it. It's a publishing house in Singapore.

Yang Chen-Ning: What's it called?

Ji Lizhen: *A Mathematician and His Mathematical Work, Selected Papers of S S Chern.*

Yang Chen-Ning: *Selected Papers* … Did Dr. Chern choose it, or did they choose it?

Ji Lizhen: It was chosen by three people: Zheng Shaoyuan, Tian Gang and Peter Wai-Kwong Li.

Yang Chen-Ning: Was this at the time after Dr. Chern died?

Ji Lizhen: When he was still alive; it was authorized by Dr. Chern.

Yang Chen-Ning: Is it one thick book, or a series of books?

Ji Lizhen: About 500 to 600 pages, or 600 to 700 pages.

Yang Chen-Ning: Wang, hand me that cube of small lattice behind you there. It was given to me by Tsinghua University on my birthday. On the three faces of the cube are written my most important work.

Ji Lizhen: That's really good. That's interesting.

Yang Chen-Ning: But my articles are short. I can neither write long articles nor textbooks. I think these 13 articles add up to less than 200 pages, I'm afraid. They are all in English, of course. They can be compiled into a book.

Wang Liping: No problem, Dr. Yang.

Ji Lizhen: Is it possible to publish the book in Chinese? Have the articles been translated into Chinese?

Yang Chen-Ning: Some of these articles have been translated into Chinese, but most of them have not because they are too professional. Publish these 13 articles, reprint them, and have someone comment on each one. Find experts to do it, because these are writings that belong to different fields. When I'm gone, you can figure this out.

Wang Liping: Actually, we can do it now.

Academic Value of Yang Chen-Ning's Anthologies and Commentaries

Yang Chen-Ning: It's not easy to find people from all 13 fields. Did I send you a copy of *Selected Papers II*? There are some photos in the *Selected Papers II*, and the thirteen papers are all there, too.

Ji Lizhen: Regarding your review, Phong from Columbia University told me last time that he liked your review very much because he also did mathematical physics. Phong, do you know him? He's Vietnamese.

Yang Chen-Ning: How do you spell Phong?

Ji Lizhen: P-H-O-N-G. 杨宏风 in Chinese, Duong H. Phong, who did string theory and was a student of Elias Stein.

Yang Chen-Ning: I don't know this guy. Where is he?

Ji Lizhen: Columbia University.

Yang Chen-Ning: At Columbia University? Is he still at Columbia University? Does he do string theory?

Ji Lizhen: He used to do harmonic analysis as a student of Elias Stein, and he also has done some string theory with Eric D' Hoker.

Yang Chen-Ning: I don't know about him. What did you say he did about me?

Yang Chen-Ning and Feynman

Ji Lizhen: A long time ago, when Feynman's book was first published, Phong said that he particularly liked your comments. He said that Feynman made everything sound very simple and easy. When he saw your remarks that everything was hard to come by, he agreed with what you said. He said Feynman gave the impression that one could do research with only a simple clue, which was actually not true. I had a very deep impression of his words. He said so decades before in the 90s. Phong came from Vietnam and did pretty good math at Columbia University.

Yang Chen-Ning: How old is he?

Ji Lizhen: He's older than me. He's 60 years old.

Yang Chen-Ning: I don't know him. I've never heard of him.

Ji Lizhen: I'll give you his link later. That's what he said to me, anyway. He said it was very nice to read your review. He thought it was very encouraging.

Yang Chen-Ning's Book Project

Yang Chen-Ning: You know, Dyson particularly liked this comment.

Ji Lizhen: I also like your review. I think ideally every expert will write about your previous work, your subsequent impact, and any recent developments, thus giving a very comprehensive picture of your work. I want to evaluate your achievement in a historical perspective. I think it would be nice to write about it this way.

Yang Chen-Ning: You may ask Shi Yu from Fudan University to

review my 13 articles. He has written several articles specifically on these 13 articles already. Let's say he has another article, in which he lists the number of times the 13 articles have been cited so far. I'll email you what he has written regarding these 13 articles.

Ji Lizhen: All right. I have another two ideas because several of the series of books edited by me at the Higher Education Press, *Panorama of Mathematics*, have recently come out. One is about Milnor, and I've selected some of his interesting articles and put them together in this book. I'd like to follow your suggestion, or we could just invite 13 people to a meeting at Tsinghua and have them talk about these 13 articles. Let's make sure they speak in simple terms, not intended for experts, and then sort out their notes based on their lectures. Because these people are usually busy to write for the book. How about doing this for your birthday? The teachers will get ready and speak to the students at Tsinghua University. You have a lot of prestige, and they would certainly agree. It's easier to write in this way and include your articles all together in the book. What do you think of this idea? It would be great if it could work.

One of the Important Differences Between Mathematics and Physics

Yang Chen-Ning: I don't know if you know that there is an essential difference between important works in physics and mathematics. Do you know this difference?

Ji Lizhen: They say that a mathematical problem can't be disproved if it's proved, right?

Yang Chen-Ning: That's right.

Ji Lizhen: This is not the case in physics.

The Eternal Beauty of Mathematics and the Value of Physics' Transition

Yang Chen-Ning: Well, I often say that if you look at mathematics in the 19th century and look at it today from a macro perspective, it is a very beautiful mountain. There are high peaks and hills, and each mountain is different in its special place. The whole scene is very beautiful. There is even a small house on the top of a mountain which may have several nice pictures in it. If this is the impression given by mathematics, then physics might give you the impression that there is a desert with several important mountains and nothing else left. In other words, in this sense, physics is very unkind to physicists. Why? Because you're guessing, and if your guess is not right, then there is no record kept in future history. And since math is not guessing, but rather, building something so beautiful that it will always be there.

Ji Lizhen: Yes, once something is proven right, it's always right.

Yang Chen-Ning: Yes, and some of the discoveries (in physics) may not have a big impact, but regarding mathematics, even later, important mathematicians and ordinary mathematicians looked at it (previous mathematical discoveries) and thought it was wonderful. We all have heard about something called a nine-point circle in middle school and all are impressed by this. What's the importance of it? I think there is no real importance. But it's not that way in physics: being either right or worthless.

Ji Lizhen: Yeah, most of the theories have been overturned.

Yang Chen-Ning: Well, for example, I think the Feynman diagram was guessed by him, so that's why people will always talk about it, not because it's something special, but because they don't understand how he made the guess, and they never will know. All in all, there's one big difference between math and physics.

Ji Lizhen: Yeah, math is different, and I also emphasize this point. We are still going to use some classic math books. Different from other scientific disciplines, the classic stuff in math is very valuable.

Yang Chen-Ning: I always say if you find a graduate student in a mathematics department at a good university and ask the student to write down the important mathematicians of the 19th century, I'm afraid everyone could write down at least 20 names, and maybe some of them could even write down 50. But if you ask a graduate student in the physics department of a good research university to write down the names of important physicists, the student wouldn't be able to write many at all. He may be able to write several names, but no more because those people have either been proven wrong later, or their works have been absorbed by other people, leaving the value of the original discovery worthless.

Ji Lizhen: Yes, history is especially ruthless, but that's the way it is. Do you want to continue with the question where we left off last time?

Yang Chen-Ning: Wang, what books have you published? For example, what research books on physics and mathematics have you published?

Wang Liping: In physics, we have translated all of Landau's textbooks (*Theoretical Physics*). There are 10 books in all.

Yang Chen-Ning: I think the textbooks were translated many years ago.

Wang Liping: Some have not been translated, and some are brand new translated versions. Have they really been translated before?

Yang Chen-Ning: I think these books came out before the "Cultural Revolution."

Wang Liping: Ours are newly published. Some previous translations should also be revised—for example, some terms should be revised. In physics, besides Landau's books, we have also translated some Nobel

Prize winners' monographs. We have a series of translated works of Nobel Prize winners in physics.

Yang Chen-Ning: You mean those who won the Nobel Prize in the past?

Wang Liping: Yes, mostly basic and textbook-type works.

Yang Chen-Ning: The World Scientific published an anthology in English. Did you translate it into Chinese?

Wang Liping: We tend to choose textbooks, such as textbooks for master's, doctoral or undergraduate students.

Yang Chen-Ning's Main Work and Landau's Ten Commandments

Yang Chen-Ning: Do you know Landau had a thing called the "ten commandments"? This is actually a bit imitative. Have you heard of it?

Ji Lizhen: Never heard of it. Ten commandments?

Yang Chen-Ning: On one Landau's birthday, and his students got a stone carved with his ten commandments.

Ji Lizhen: Because the Ten Commandments God gave Moses were carved on a stone.

Yang Chen-Ning: Have you heard of it?

Ji Lizhen: I don't know anything about Landau.

Yang Chen-Ning: I'll give you Shi Yu's article. He has already referred to it. There are articles discussing it.

Landau's Toughness and Acrimony

Ji Lizhen: Yeah, yeah. Landau is very mean to people.

Yang Chen-Ning: After Landau died, some of his students wrote a book to commemorate him. These students, of course, admired him very much, but the Landau style described here is not desirable, esp. from Chinese perspective.

Ji Lizhen: Yeah, he was very domineering.

Yang Chen-Ning: He was so aggressive with his students that he often made some of them look down on themselves. For example, about once a week he would gather the students for discussion. When one of the students spoke up, he swore at the student and said something along the lines of, when you were in kindergarten, didn't your mother teach you how to behave?

Ji Lizhen: I also heard of his ten physics courses under the name Landau–Lifshitz. Later, Landau said that every idea in them was his, and that every mistake was Lifshitz's. I thought it was very harsh.

Yang Chen-Ning: I have the impression that Gelfand was a bit like Landau when he was in the Soviet Union.

Ji Lizhen: Really? Yeah, so I guess maybe it's good in a way because they had mentored a lot of good students.

Yang Chen-Ning: Do you know Hao Bailin? Actually, he works at the Institute of Theoretical Physics, but he doesn't come here very often now. He's at Fudan University. He's a very good physicist. He's developed a new field in the last 20 years, and he's been very successful in it. Well, there's a taxonomy in biology, and he asked about two decades ago that whether cells could be classified as well. Not a taxonomy of organisms, a taxonomy of cells. The classification of a cell is much more complicated than that of organisms. Because cells are small and simple, there are many variations and intricacies in them. He developed this field, which is also being done in the West, and in this field, he is now a very important man in the world. Also, he's kind of straightforward. He's a student of Landau's. If you want to ask about Landau, you can visit him. He was probably a student of Landau's who didn't graduate, that is, Landau passed away before he graduated. Landau got hit by a car.

Ji Lizhen: Yeah, he came to an abrupt end. Okay, let's start over.

Collaboration of Fermi and Yang Chen-Ning

Yang Chen-Ning: I'll tell you one thing. There's an article here about Fermi.

Ji Lizhen: Yes, I have read it.

Yang Chen-Ning: So you have read that article?

Ji Lizhen: Yes, I have.

Yang Chen-Ning: What I think is important about that article is what I discussed earlier about Fermi's way of being a man.

Ji Lizhen: Yeah, I remember the article quite well. You two wanted to develop a criterion to measure something, and Fermi spent several periods of time working with you on it, and I thought it was amazing.

Yang Chen-Ning: His principles were quite in line with the traditional Chinese culture, while the United States was going in another direction entirely. In fact, this is a very important topic. Namely, the Chinese cultural tradition requires people to show introspection toward themselves three times every day. People should be saints on the inside and kings on the outside. The United States is focused on developing more outwardly. The United States doesn't emphasize the principles of being a good man. It focuses more on what is expressed outwardly.

Ji Lizhen: Yeah, that's what Feynman was like.

Yang Chen-Ning: That's what Feynman was like, and Landau was like that to some extent, allowing people to develop that way. I don't think we can say which one is good or which is bad, but we need to understand the advantages and disadvantages of each approach. There's actually profound knowledge.

Importance of String Theory

Ji Lizhen: We reached the third question in the last interview, and now let's come to the fourth question about the depth and breadth of knowledge. You actually talked about this yesterday. I think everyone thinks Einstein's work is very profound, but the question is, can you give some examples? Sometimes there's something very hot for a while, but then it's not so popular afterwards. Why is that? I think string theory is very hot right now in physics and mathematical physics. What do you think about the theory? Of course, only time can tell how it is going to play out. Because I had a string theory session here last August, which I didn't understand well, and I was listening to it intently. I was kind of surprised that it didn't seem to give me any new ideas. What do you think of this kind of problem, that is, what is real, profound and theoretical knowledge in this field?

Yang Chen-Ning: As a matter of fact, I think everyone who is interested in the development of mathematics and physics has a similar question. This string theory, if you want to describe it in simple terms, what does it do? I think it is meaningful because the most important question that physics hasn't solved is quantum gravity, and gravity still hasn't been quantized yet. There are four forces in the world: gravity, weak interaction, electromagnetism, and strong interaction, and man knows much about three of these forces now.

Ji Lizhen: Yeah, there's consensus in understanding.

Yang Chen-Ning: Quantum theory is now called the Standard Model, but it cannot explain gravity from the very beginning. Now you ask me about it, there's still no right way to approach it. But despite that premise, it turns out that the other three that we now know are very beautiful mathematical structures having to do with geometry, specifically within the context of fiber bundles. So there was a belief that for unsolved fundamental problems in physics, there had to be some new mathematical structure, which became an exploratory direction. If you ask for my thoughts on that, I think most people who study physics, also people who are related to mathematics, and mathematicians related to physics, like Atiyah, think that this is a creed, something that you can believe in. But how do you put it into practice? One way to do that, which is what string theory is doing now, is to find, at some point, a nice geometric structure in ten dimensions. This is currently very lively and has been going on for 30 or 40 years. Some good results have come out for which mathematicians admire very much.

Ji Lizhen: Yeah, it's great.

Yang Chen-Ning: For example, Atiyah is very impressed. Not just Atiyah, a lot of other people are very impressed, too. In this way, because there are some high-dimensional structures that are very complicated and have beautiful math in them, it attracts young people. There are probably 10,000 people around the world working on string theory right now. But what is happening now is different from traditional physics; it has nothing to do with experiments, nothing to do with them at all. It deals entirely with mathematics …

Ji Lizhen: That's impossible to verify.

Conjectures of String Theory Cannot Be Tested Experimentally

Yang Chen-Ning: Not only was string theory untestable, but also he couldn't even come up with an experiment to verify his theory and prove if he was right or wrong. So, it's like he was there, and he had built something that had nothing to do with the experiment. So far, of course, they'd like to prove it, but they can't. So, there's a very interesting phenomenon: these 10,000 people all go to the physics department, not to the mathematics department because the mathematics department doesn't want them. Why don't they go to the mathematics department? Because string theory isn't mathematically rigorous. So, the mathematics department has to talk about it because after they hire somebody, their faculty can't get papers published in the traditional mathematical journals. Instead, they end up getting their papers published in physics journals. Because physics is not so rigorous, physics allows for guesswork, so good physics journals accept such papers, and many new journals are being published. Given that these scholars' papers can be successfully published in physics journals, the physics departments accept them, so those 10,000 people work basically in physics departments all over the world instead of mathematics. If you ask me how it will develop five or ten years from now, I don't see how that's going to change. Of course, in five or ten years, the number of people will continue to increase, so there will be some impact on the development of physics and mathematics as a whole. No one dares to claim what the impact will be in the future. The most important person here now, as you probably know, is Edward Witten. Witten has a great influence on mathematics.

Ji Lizhen: Very big influence.

Witten and Langlands Program

Yang Chen-Ning: I think this guy is a genius. I hear he's working on the Langlands Program now. But I don't know whether it has worked out.

Ji Lizhen: His Langlands program is different from the original one, very different. Basically, there's no relation.

Yang Chen-Ning: You said it wasn't about …

Ji Lizhen: ... about number theory.

Yang Chen-Ning: Is it not exactly the same as in mathematics?

Ji Lizhen: It's very different.

Yang Chen-Ning: You mean some of the articles Witten wrote?

Ji Lizhen: Yeah, the articles he wrote were different.

Yang Chen-Ning: So, the development of string theory is a strange phenomenon in 200 years of physics development.

Yang Chen-Ning's Philosophical and Religious Views

Ji Lizhen: The following question is about the relationship between science and humanity, and this question is also related to another question. I read a book in which you were interviewed by the Public TV Station around the turn of the century in 2000, and you talked about science, philosophy and religion. Now we're going to link these up: science and humanities, science and philosophy, and religion. What do you think? I'm particularly interested in what you think of religion.

Yang Chen-Ning: I'm not a person who studies philosophy, let alone religion. I can only give you my layman's general opinion. I think basically no matter religion, philosophy or science, they all have the same ultimate goal: to understand the relationship between man and nature. But the emphasis is different, the methods are different, so the established traditions are different. They are changing. What science does today is a little bit different from what science did 100 years ago. In particular, philosophy has changed, and religion has changed a lot. I think in general, science is gradually shrinking into the realm of philosophy, and philosophy is shrinking into the realm of religion, so these three fields—science, philosophy and religion— are changing, but at different times. I think this is quite natural, and I don't think that if things go on like this, philosophy and religion will be overwhelmed by science in the future. In other words, science ... There's an article in this section called The Future of Physics. It was from 1961, and there was a big celebration for the 100th anniversary of the founding of the Massachusetts Institute of Technology. Many small discussions were held at the celebration. Among them, there was a panel discussion called the Future of Physics, which was made up of four people. The four men were Feynman, a British man named John Douglas Cockcroft, a Nobel Prize-winning experimental physicist who had invented the electrostatic accelerator, Rudolf E. Peierls, who was a very important theoretical physicist in England and also the first person to estimate what the critical mass of uranium is, and me. Peierls's work had a big impact at the time, because between 1948 and 1950, all physicists knew that it was possible to make a nuclear bomb out of uranium. How much uranium was needed to make one nuclear bomb? There was a minimum amount of uranium that was needed to make a nuclear bomb, and this was called the critical path, and this critical path was first estimated by Peierls. His estimate was wrong, but it was a very famous experiment in history, and he made a fundamental and important contribution to physics. The most important thing that was noticed in the discussion between the four of us was that Feynman had a completely different attitude than I did. I discussed the same topic two years ago, so I'll be sending you that.

Ji Lizhen: Okay. It's in here, right?

Yang Chen-Ning: It's not here. My article is here, and the Feynman's article is also published now, so I will send them all to you. The main thing is that I'm not like him at all. He's very optimistic about the future. He thinks that everything will be solved soon. But I think it is impossible. I think science will have worldly success, but it will never catch up with nature, which is much more complicated. My basic argument is that there are a finite number of scientific successes, but the complexity of the universe is infinite. In fact, this can be applied to mathematics. I believe that the mathematics of beauty is infinite. It is impossible to know all the mathematics of beauty. Why is that? As biologists have discovered, the list of really weird things about biology is endless. I think there are an infinite number of strange, unimaginably beautiful phenomena in the biological world. And this mathematical structure, its beauty, is infinite. Therefore, I believe that mathematics is infinite in knowledge.

The Most Important Thing in Life

Ji Lizhen: Right, Okay, now let's move on from this question to the next. Because now that we're talking about religion, so I wonder what is the most important thing in a person's life? I find that is a difficult question to answer. We used to learn about this when we were in college, which I think was naive of course. Now that I'm in my fifties, I sometimes do wonder why I am alive. I'd like to hear what you think about it because you've seen a lot, and you've experienced a lot. What do you think about this from your point of view? For example, this question is related to several questions. What is really the most important thing for a great scientist, and what legacy does he or she want to leave?

Yang Chen-Ning: What did you say?

Ji Lizhen: It's a legacy. I mean a legacy. What do you want to leave behind? What is the most important thing to them?

Yang Chen-Ning: I think the structure of mathematics by itself is one thing, and it's another thing for humans to know the structure of mathematics, and of course, the two are mutually [related], but they're also quite different. In other words, when there were no humans in the world, there was already a structure of mathematics. How can I prove it? Because when there were no people in the world, there was radiation, and that radiation conformed to Maxwell's equation.

Ji Lizhen: Yeah, it's a natural thing.

Yang Chen-Ning: So, Maxwell's equations were already there. The mathematical structure was already there. It had nothing to do with whether there were people or not.

How do humans understand this structure? It's a very curious phenomenon. Because if you look at how different people are sensitive to this structure, take Gauss or Galois for instance, they are sensitive to mathematics, and this is what they are most sensitive to. In fact, I think, I'm afraid they could feel it even at the age of three or four. This doesn't mean there was a record, it just means that Gauss was sensitive from a very young age. Dyson, too, said that he seemed to know even when he couldn't speak clearly, he just thought it over and found out that when he added 1/2 + 1/4 + 1/8, he had to add 1. There was a difference between his brain and the brains of ordinary people. Why was there this difference?

Ji Lizhen: Talent, right?

Yang Chen-Ning: I think that this mystery and issues related to religion will never be made clear. You can know a lot about them, but there is no way to understand these strange phenomena completely. So that's why I basically disagree with Feynman about the future of physics.

Ji Lizhen: Well, I'd like to be specific. For example, in China these days, people who are relatively wealthy seem to value money more than Americans do.

Yang Chen-Ning: How is China more than America?

Ji Lizhen: I feel like everything is about money. You've been through a lot of experiences. What do you think is most important in a person's life? I don't mean that in an abstract way. For example, for a general physicist or mathematician, is the purpose of life really to find a better job or to get a professorship after graduation? I guess not. What's really important in life? I'm asking you a more concrete question. What do you think?

Meaning of Life and Deng Xiaoping

Yang Chen-Ning: I think the great contribution of Deng Xiaoping is that he thought capitalism was not all bad, so he wanted to bring in the good things about capitalism, called socialism with Chinese characteristics, and the basic point of socialism with Chinese characteristics was to use capitalism, especially in the 1990s. I've been reading a book. I don't know if you've read it.

Ji Lizhen: What book?

Yang Chen-Ning: Someone gave me a Chinese translation of the book by Henry M. Paulson. This is a book you'd better read.

Ji Lizhen: What's it about? Paulson's book. What's it about?

Yang Chen-Ning: It's called *Dealing with China*.

Ji Lizhen: Oh, I don't know about that one. It's about the changes in China. Okay.

Yang Chen-Ning: He's the CEO (chief executive officer) of Goldman Sachs. It's been translated into Chinese now. I'm reading it in Chinese. Someone sent me a copy, and it's good; the original one is in English. There is one part that is very interesting. It talks about Deng Xiaoping's southern tour. If you ask about his southern tour, it is actually what Deng Xiaoping intended to say that we must introduce capitalism (market economy), and that the capitalism (market economy) which we want to control can be introduced. I think he had the idea, and he had the ability to implement it. That's what's great about him. The impact was not just on the Chinese people, but on the whole world.

Ji Lizhen: Yeah, it affected the whole world. So, it is difficult to answer to what extent Deng Xiaoping influenced the country and the whole world. So, what do you think is more important in a person's life? What do you suggest? I think for ordinary people, many pursue fame and wealth, but the question is, what is left after fame and wealth, and I think this is related to religion.

Yang Chen-Ning: What mathematics is talking about, what physicists are talking about, what politicians are talking about are getting more and more complicated.

Ji Lizhen: I'm asking a very simple question.

Yang Chen-Ning: But there is one thing in common. Mathematicians see problems more clearly. Deng Xiaoping saw problems less clearly and in more complicated ways. But success in all these areas has one common element: the ability to go beyond the complexities of the present and grasp the problem at large. It's the same in all of these areas, and it's about being able to capture that essence. Why do some people do math but not physics, and why do some people do physics but not math? I think it has something to do with the software in their heads. Let's go to dinner.

Understanding the Meaning of Life

Ji Lizhen: So, you say the same is true of life, to grasp the essence?

Yang Chen-Ning: Yes. Do you read Jin Yong's kung fu novels?

Ji Lizhen: Yes, I've read them all.

Yang Chen-Ning: When I was a kid …

Ji Lizhen: You didn't think they were engaging?

Yang Chen-Ning: I like to read old-style kung fu novels, one which you probably don't know is called *The Seven Heroes and Five Righteous*, which was very famous at that time, and it was called *The Seven Swords and Thirteen Heroes* back then. But when I entered university later, I could not bear to read it; I thought that I could not bear to read things that were so imaginary.

Ji Lizhen: I read Jin Yong's novels, and they were mainly about love stories. I think they are better.

Yang Chen-Ning: Jin Yong sent me a few of his novels. I started reading them and couldn't finish reading.

(At noon, Yang Chen-Ning invited Ji Lizhen and Wang Liping to have a potluck in Jingzhai, Tsinghua University. During the meal, the interview continued.)

Ji Lizhen: Dr. Yang, the question I asked you in the car is like this. People are talking about the explosion of knowledge. For example, there are a lot more articles than there were in previous decades. What do you think of that? I guess there's probably not much that really matters.

Knowledge Explosion Is Not a Real Explosion

Yang Chen-Ning: Yes, I totally agree. Some of them are correct, but I think many of the details are incorrect. I think it's true that there was an explosion about 13 billion years ago. It is also generally true that there were some phenomena that came out of the Big Bang, including light elements at the beginning, and gradually all the elements of matter were forged through explosions and subsequent reactions, but there is still a lot that is not understood thoroughly here.

Ji Lizhen: I'm talking about the knowledge explosion.

Yang Chen-Ning: Knowledge?

Ji Lizhen: The knowledge explosion. Because people are saying we're publishing a lot, like we're doing more data now than we've done in the last thousand years combined. In physics there are also more and more articles and books.

Yang Chen-Ning: When do mathematicians think the explosion happened?

Ji Lizhen: No, they just said there are a lot of journals, there are a lot of books, and now I'm asking what is really essential. How do you judge an article or a book, and what's really more important? Of course, time will judge everything, but the problem is that sometimes I can't wait 100 years to look back. Does the same go for your publishing? There are a lot of magazines and a lot of books, and I don't think most of them are really that important. Let me ask you a specific question. When you look at the Nobel Prize in Physics, it has been a long time coming. In your opinion, who among the Nobel Prize candidates or recipients in physics have truly outstanding work? Are there some Nobel Prize winners who didn't do well enough in the course of their (later) work?

The Great Breakthrough in Physics and the Greatest Physicist

Yang Chen-Ning: I think it's very clear that, in terms of physics, the first and most important work was undoubtedly by Newton, and it wasn't just physics that was impacted.

Ji Lizhen: Mathematics?

Yang Chen-Ning: It's not just mathematics. It's the first time, through his work, that he made people realize that there are laws in nature that can be described by accurate mathematics. There was no such idea before.

Ji Lizhen: The first time, right?

Yang Chen-Ning: The first time. There have been some before, like Johannes Kepler. Kepler had some of this type of idea, but he was not accurate enough. He was accurate to a certain extent, then came Pierre-Simon Laplace and Joseph-Louis Lagrange, claiming that the orbit of the planet could be calculated accurately (I don't know how accurate it was in the 18th century). I think it was accurate to one ten thousandth, which at the time was unprecedented for human beings. Do you know how accurate it is now?

Ji Lizhen: I don't know.

Yang Chen-Ning: I don't know astronomically, of course, but it is very accurate now. But physics is even more accurate. Now it's up to one billionth.

Ji Lizhen: That's very accurate.

Yang Chen-Ning: It's renormalization, which can be calculated, and compared with experiments, so it's now up to one billionth accuracy.

Ji Lizhen: That's great. Newton was the first. Who was the second?

Yang Chen-Ning: I think the second one would be James Clerk Maxwell.

Ji Lizhen: Oh, Maxwell.

Yang Chen-Ning: Of course, it's not Maxwell's contribution alone, it's Maxwell's research on Michael Faraday. Faraday didn't study mathematics, so Faraday's writing didn't have any formulas, but he had intuition and insight. Maxwell admired Faraday very much. Maxwell said that Faraday was actually a first-class mathematician, that is, he had good intuition, so Maxwell summed it up from Faraday's article and wrote it into a formula, which was Maxwell's equation. The influence of this has been very great.

By the 20th century, I think there were a number of important things. One, of course, was Einstein, and one was quantum mechanics.

Ji Lizhen: Was it Planck? Or who did that?

Yang Chen-Ning: Quantum mechanics started slowly, first by Planck in 1900, next Einstein in 1905, then Bohr in 1912 and 1913, and by 1925 or so there were just three: Heisenberg, Dirac and Schrödinger.

Ji Lizhen: So that's it, Dirac, Einstein, Heisenberg and Schrödinger?

Four, any more? Actually, in physics, since Newton, there have been only a few major people, and there's one more.

Yang Chen-Ning: Pardon?

Top Five Physicists in the Field of Quantum Mechanics

Ji Lizhen: You just told us that physics started with Newton followed by Maxwell, and then quantum mechanics, in which you gave these examples: Einstein, Dirac, Heisenberg, Schrödinger and Bohr, right?

Yang Chen-Ning: Yeah.

Ji Lizhen: It's mainly these five people.

Yang Chen-Ning: Each of them went in different directions to create quantum mechanics. I think there are also some phenomena in mathematics. For example, when you teach students Lie groups today, you never start in the direction of Sophus Lie, because that direction is extremely difficult.

Ji Lizhen: Yeah, Sophus Lie originally intended to solve equations.

Yang Chen-Ning: In fact, when I wanted to learn Lie groups in the 1950s, these textbooks on Lie groups still started with Lie's transformation groups. That path was quite complicated. No textbooks do that now.

Ji Lizhen: Yeah, it's more abstract now.

Yang Chen-Ning: Right now, it probably starts with Lie algebra.

Ji Lizhen: After Schrödinger and Heisenberg, who are the main people in physics making particularly outstanding contributions? For example, 100 years from now, who may still be remembered for their contributions?

Yang Chen-Ning: The beginning of quantum mechanics is equivalent to taking our understanding of the physical world in a whole new direction, so all physics is starting over. This impact is huge because of the semiconductor. If there had been no semiconductors, there could not have been a network, and there wouldn't be today's mobile phone.

Ji Lizhen: Right. On the other hand, because I'm not very familiar with physics, sometimes when I come across the name Steven Weinberg, I notice that quite a few physicists have won the Nobel Prize. I've heard a lot about Feynman, about you and Lee Tsung-Dao. I've heard about Yang Chen-Ning and Lee Tsung-Dao from very early on, but other Nobel Prize winners are kind of vague to me. Feynman's PR efforts had a significant and widespread impact and someone else wrote "The Unreasonable Effectiveness of Mathematics in the Natural Sciences". Who was that? Wigner, or whoever wrote it.

Important Directions in Physics

Yang Chen-Ning: I think the future development of quantum mechanics you just mentioned is one direction. The other direction of development, which I'm afraid those who study mathematics have not noticed much, is called solid-state physics.

Ji Lizhen: Solid-state physics? No, we don't know about that.

Yang Chen-Ning: This solid-state physics is very important for applications.

Ji Lizhen: Yes. Who are the main representatives?

Solid-State Physics and Semiconductors

Yang Chen-Ning: Like semiconductors. Why are semiconductors so important?

Ji Lizhen: Because all electronic products depend on these?

Yang Chen-Ning: That's right. By the 1920s, they could make radios. When I was in high school, everyone would make a little radio receiver out of a set of crystals, so a lot of people in high school were building these in their homes. What was its main principle? Some wires were connected, and you could control whether they were connected or not. We all know that there is an electric switch that can be turned on and off, with just a twist. But when the special wire was in use, it had a kind of switch in it. The switch was controlled by electricity, not by hand. So, by the 1930s, everyone knew that using electricity to turn on or off to control a wire was a key skill. As a result, in the 1950s, when semiconductors were discovered, the developers personally understood immediately that a wire could be controlled by a semiconductor. What is a semiconductor? Well, something is not a conductor, but you control it and affect it with electricity, then it can become a conductor. Affect it again, and it can become a non-conductor.

Ji Lizhen: Right, it's not mechanical.

Yang Chen-Ning: This understanding is the beginning of industry in modern times. Because at that time, in the 1930s, our middle school students all used crystals to make radios, and in the 1940s, they all used vacuum tubes. You probably haven't ever seen vacuum tubes.

Ji Lizhen: No, I haven't. I've just heard about them. I don't know what vacuum tubes are.

Yang Chen-Ning: The size of a vacuum tube was reduced to a small point when semiconductors came out, and that's where modern computers came in.

Ji Lizhen: Are there semiconductors in mobile phones these days, or is there something better?

Yang Chen-Ning: Everything is still semiconductors. It's just that semiconductors are getting smaller and smaller, and they're still getting smaller.

Ji Lizhen: The discovery of semiconductors was a very important milestone in physics.

Yang Chen-Ning: It had a huge impact on applications. I met the former president of Fudan University, Xie Xide, who is famous for her work in semiconductors. The originators of China's semiconductors are Huang Kun and Xie Xide. Huang Kun was in Beijing and Xie Xide was in Shanghai, but in the 1950s they co-organized seminars and introduced the physics of semiconductors to China for the first time. Today, all the people engaged in semiconductor research and practice in China are their disciples and followers.

Ji Lizhen: They trained them. Okay, moving on, what's so important about physics that it had such a big impact?

Superconductivity in Solid-State Physics

Yang Chen-Ning: Another particularly important discovery, which also belongs to solid-state physics, is superconductivity and superfluidity. Do you know what superconductivity means and how it was discovered?

Ji Lizhen: I don't know.

Yang Chen-Ning: In the 19th century, electricity and electrification were clearly studied. You must have heard of Ohm's law. At that time, especially because of industrial applications, the electrical conductivity of different materials was an important topic. Copper, for example, is highly conductive.

Ji Lizhen: Gold's conductivity is high, too.

Yang Chen-Ning: Iron is not as high as copper, so today wires are made of copper, not iron. Then, people found that when the temperature is different, the conductivity of the same wire also changes with the temperature. Around 1910, there was a lab in Leiden, a very small city, but very industrial. In this lab, there was an experimental physicist named Kamerlingh Onnes, and he developed a technology that could cool things better than anything else. In fact, for ten years after that, no other place in the world could reach the temperature that he did. What was the temperature?

Ji Lizhen: Absolutely zero, right? Almost?

Yang Chen-Ning: Several degrees Kelvin (thermodynamic temperature). In the great contributions we have just talked about, we didn't mention thermodynamics, which is, of course, a huge contribution.

Ji Lizhen: It was Ludwig E. Boltzmann, wasn't it?

Yang Chen-Ning: No. Thermodynamics was done by several people who, I think, for a mathematician are probably not so famous. Well, the steam engine had already been built in the 18th century, so it was a matter of desperately studying how to make the most efficient steam engine. By the middle of the 19th century, they figured out that the efficiency of the steam engine couldn't be increased indefinitely; it was finite, and they worked hard on that, then thermodynamic laws came out. These laws of thermodynamics will tell you that there is a limit to the efficiency of a steam engine, and that limit is related to the temperature at which it is used. This gives rise to the idea of absolute temperature, which is a first-class contribution to the principle of physics. Boltzmann came later than this. Boltzmann explained the causes of thermodynamics as molecular, so that's also called the concept of statistical mechanics. If you ask about the three most important contributions of the 19th century, one is electromagnetism, one is thermodynamics, and one is statistical mechanics; those are the three most important contributions of the 19th century. And then in the early 20th century, while studying the electrical conductivity of this material, Kamerlingh Onnes discovered a new superconductivity when he cooled the material to a temperature near absolute 3 or 4 degrees, which is −270 degree Celsius. He made it, but no one else could. Then, he took different materials, such as copper, lead, and tin, put them in a cold place, and measured their electrical conductivity, so he published a lot of articles. He suddenly discovered (and I think he was the first to discover), I can't remember which material it was, that its electrical conductivity at a temperature below several degrees still flowed. He used a coil, then electricity was applied to that coil, making its temperature drop … Let's say he took a capacitor, and a capacitor has two poles, one positive and one negative, and then he took a coil. If you connect the coil, it has a current. He dropped the coil down to a very low temperature, and he found that after the current flows, even if you don't add energy to it, it still always flows. In fact, this is beyond imagination, because …

Wang Liping: It doesn't change direction. Is it a directional flow?

Yang Chen-Ning: It continues to flow in one direction. Usually, when you turn off the power supply, it automatically becomes weaker, and then it goes away.

Wang Liping: Is it weak or flowing in the opposite direction?

Yang Chen-Ning: It goes in one direction, then it gets weaker and weaker, and eventually it will disappear.

Ji Lizhen: You just said it kept flowing?

Yang Chen-Ning: He didn't believe it was always there, so he did an experiment and went back to see it months later. It was still there.

Ji Lizhen: Yes, it's superconducting.

Yang Chen-Ning: This is called superconductivity. This was a great new discovery; it had not been understood at that time.

Ji Lizhen: Is this a relatively new concept? It seems that Chern Shiing-Shen's son-in-law is also doing this, right?

Yang Chen-Ning: Yes, the story went like this, then all the people learned about this superconductivity. Of course, this thing must have had practical applications, because you know that when electricity is sent to people from various power plants, it loses some energy. If you use a superconducting wire, can you avoid these losses? It had big implications for industry.

Ji Lizhen: But how do you keep the temperature so low?

Yang Chen-Ning: You have to spend money to bring it down to a low temperature, and spending a little money to bring it down to a low temperature can reduce consumption, so it needs to be balanced. Of course, it is difficult to achieve such a low temperature, so this is only a distant idea. No one can implement it fully, but this is the future direction, so we should continue to research and see: can a little higher temperature have superconductivity? In the 20th century, it became a very popular field of knowledge, and everyone was working on it like crazy. As a result, in the 1950s, it was discovered that the Kamerlingh Onnes had found a material that superconducted at a temperature of several degrees (absolute temperature), and then the temperature was gradually increased, and by the 1950s it was discovered that the temperature could reach 23 degrees (absolute temperature). If a person finds a material that can become superconducting at 23.3 degrees, he will immediately become an authority and then …

Ji Lizhen: Do you mean 23 degrees K or our 23 degrees?

Yang Chen-Ning: K.

Ji Lizhen: How much is K equivalent to when being converted to our Celsius and Fahrenheit?

Yang Chen-Ning: 0 K is –273 degrees, so 23 degrees K is –250 degrees.

Ji Lizhen: I mean when you convert Celsius to Fahrenheit, for example, you have to multiply it by something.

Yang Chen-Ning: Physicists don't use Fahrenheit, they use Celsius, and –273 degrees in Celsius is called absolute zero.

By the 1950s, it was thought that it couldn't get any higher than 23 degrees. At that time, it was only around 23 degrees, and a little bit made a difference. Paul Chu, who is Dr. Chern's son-in-law, along with a guy named Bernd T. Matthias (Matthias was a great expert on this, and Paul Chu was a student of Matthias's), did very well. He was an expert on this in the 1950s, 1960s and 1970s, but all were under 23 degrees.

Ji Lizhen: There's no improvement yet, is it?

Yang Chen-Ning: 23 degrees is too low to achieve in a lab, so it wasn't used at that time. And it had been written that it was impossible to go above 23 degrees. Then, all of a sudden in 1986, there were two Germans in Zurich, Switzerland, who published a paper. They had found a type of insulator, insulate, which was an insulator at ordinary temperatures, but became a superconductor at absolute temperatures of 30 degrees. Why did this matter cause a sensation worldwide? After World War II, there were two things that made the biggest splash in physics, two things that suddenly shocked the whole world of physics. The first was Wu Chien-Shiung's parity non-conservation in 1957, which shocked the whole physics world. The second was that in 1986 and 1987 the superconductor suddenly broke the 23-degree limit, which had been thought impossible for many years.

At that time, the whole world was working on this, and there was a conference. I remembered there were thousands of people working in this field. Paul Chu worked together with his Ph.D. student, Wu Maokun. They used something like what Zurich had done, made some adjustments, and the temperature reached 90 degrees.

Ji Lizhen: That's pretty high, 90 degrees.

Yang Chen-Ning: To go from 20 degrees to 90 degrees is not just an increase in number, it has very important implications. Why is that? Because to reach a temperature of over 20 degrees, you need something cold, and that cold thing is liquid helium. Liquid helium is very expensive. If you get to 90 degrees, you don't need liquid helium; you need liquid nitrogen. If you compare the price of liquid nitrogen to that of liquid helium, it's like the price difference between milk and XO cognac. In other words, when it comes to liquid nitrogen, it's ready to be made because it's easy and cheap to make. With liquid helium, it's a big cost and too expensive. So, all of a sudden it becomes usable, and it is still used today. Now the temperature has risen again, to more than 100 degrees, but still they're using liquid nitrogen.

Another phenomenon, similar to this one, is called superfluidity. Superfluidity was discovered in the 1930s, and it was first discovered when liquid helium was found. When superfluidity was passed through a tube, it didn't have viscosity. Normally, when a liquid goes through a tube, it loses some energy. He found that at low enough temperatures, liquid helium loses its resistance. This is called superfluidity. So, this is superfluidity and superconductivity …

Ji Lizhen: They are kind of similar.

Superconductivity and Mathematics

Yang Chen-Ning: This is an important question. In fact, one of my 13 articles was published in 1961, and I named it "Off Diagonal Long-Range Order." I have a general understanding of superconductivity and superfluidity. There is a mathematical theorem in it, and I can tell you how it works. It's like this, it's a Hilbert space, and it's based on integers going from negative infinity to positive infinity.

Ji Lizhen: They are integers? A set from negative infinity to positive infinity, right?

Yang Chen-Ning: This one is one-dimensional. And then you can make it two-dimensional, and three-dimensional, which is a two-dimensional plane, and you label a Hilbert space on it. The function up here (you asked about symmetry) could be symmetric, or antisymmetric, this x and y. So, everything that's symmetric is a Hilbert space, and everything that's antisymmetric is also a Hilbert space. This phenomenon is said to be a wave function of matter particles; for some particles, its wave function is symmetric, and these particles are called Bose–Einstein particles. The others are called Fermi–Dirac particles. The idea started with Bose and Einstein in 1924.

Ji Lizhen: Condensation. That's what you told me last week, wasn't it?

Yang Chen-Ning: It's basic. It's very basic. You have got to an n-dimensional wave function that is not a two-dimensional space. Its symmetry, so far as we know, is of two types. One is symmetric and one is antisymmetric. I've figured that out, too. There's a theorem here, and the mathematical meaning of the theorem is this. I'm very proud that I proved this theorem. It's a mathematical theorem, but if I have to explain this mathematical theorem to a mathematician, wait and let me think about it. I'll have to think about that, and when I figure it out, I can email you a precise mathematical description of the theorem.

Ji Lizhen: Okay, thank you.

Yang Chen-Ning: The proof is not too complicated. By not too complicated, I mean that any freshman who has a good score in mathematics can understand it. And I only proved the case of one parameter called n, $n = 1$ and $n = 2$. And then I made guesses when $n = 3, 4, 5$, but I couldn't prove it. There are still people working on this, and I'll tell you the details. Its equations are related to the N-dimensional Hilbert space of symmetric particles, what is the maximum symmetry of it, and that kind of thing. This kind of problem is very closely related to superconductivity. I'll have to figure it out. Recently Lin Kailiang told me … By the way, how do you know Lin Kailiang?

Ji Lizhen: Because he is compiling a series of books called *Mathematics and Humanities*.

Yang Chen-Ning: Is it *Mathematics, Science, History and Culture*?

Ji Lizhen: *Mathematics, Science, History and Culture* is from Chinese Taiwan. Our series is called *Mathematics and Humanities*.

Yang Chen-Ning: Where was it published?

Ji Lizhen: It was published by the Higher Education Press.

Yang Chen-Ning: Is it called *Mathematics and Humanities*?

Wang Liping: Yes, it's a series of books. It's not a journal. It's not a magazine.

Yang Chen-Ning: What's in it, for example?

Wang Liping: It tells some stories about mathematics and mathematicians.

Ji Lizhen: One of the books we gave you last year was included in that series. It's about Southwest Associated University in Kunming.

Yang Chen-Ning: Is it about the Riemann conjecture? The guy with the surname Lu.

Wang Liping: Lu Changhai.

Yang Chen-Ning: Yes, he wrote a book on the Riemann conjecture.

Wang Liping: It was not included in the series.

Yang Chen-Ning: Wasn't the book published by you?

Wang Liping: Not us.

Yang Chen-Ning: It's very well written.

Wang Liping: He is now helping us write a book about gravitational waves, i.e., gravitational field.

Yang Chen-Ning: Does he have anything to do with that?

Wang Liping: He's going to write a popular science book about gravitational waves.

Yang Chen-Ning: About gravitational waves?

Wang Liping: Yes, about gravitational waves. I can't remember whether Riemann's book is published by the Tsinghua University Press or Science Press.

Yang Chen-Ning: What is your book title? *Mathematics and Humanities* is a series of books, many of which have been published.

Wang Liping: Almost twenty.

Yang Chen-Ning: For example, you can tell me about some topics, and I'll listen.

Wang Liping: For example, there is a special section on female mathematicians.

Yang Chen-Ning: Female mathematicians. Is it written in Chinese?

Wang Liping: Yes, in Chinese. For example, there's one devoted to math and music.

Yang Chen-Ning: Is there one on female physicists?

Wang Liping: No.

Yang Chen-Ning: Recently, someone gave me a book in English about women scientists. Two of them are Chinese, Wu Chien-Shiung and Wu Sau-Lan.

Wang Liping: Wu Sau-Lan—is it not about Chang Sun-Yung?

Ji Lizhen: Chang Sun-Yung was a mathematician.

Yang Chen-Ning: Chang Sun-Yung is the one from Princeton?

Wang Liping: Yes.

Yang Chen-Ning: I haven't met her. What field is she in?

Ji Lizhen: She does geometric analysis, PDE (partial differential equations).

The Second Handshake and Wu Chien-Shiung

Ji Lizhen: I heard that Wu Chien-Shiung has something to do with the novel *The Second Handshake*. Was she the protocol of the novel?

Yang Chen-Ning: *The Second Handshake* was written by someone who didn't know her and made up a novel after reading some articles. At that time, because there were few such novels in China, people copied them by hand, so it became a very popular thing at that time. After I read it, I didn't think it was very good.

Ji Lizhen: Nothing to do with Wu Chien-Shiung?

Yang Chen-Ning: Basically, it is nothing related, because the author doesn't know physics. Wu Chien-Shiung's work is very important. Didn't I tell you last time?

Ji Lizhen: That's right. I don't think she ever won a Nobel Prize.

Yang Chen-Ning: No, but she should. I think in the future, there will be people who will write about the reason why they didn't give it to Wu Chien-Shiung.

Wang Liping: Mr. Ji, don't you have a question about why the Nobel Prize is not given to mathematicians?

Ji Lizhen: Yes. Why do you think there is no Nobel Prize for mathematics? What do you think about it?

No Nobel Prize for Mathematics

Yang Chen-Ning: There is a rumor, but some people say that the rumor is not true. They say it is because of Mittag-Leffler… but I think these messy facts may not be necessary. Because Nobel was an industrialist, I think he was not as interested in pure mathematics as he was in practical applications, so he had physics, chemistry and biology, which are all related to practice.

Ji Lizhen: Yeah, literature too.

Yang Chen-Ning: He didn't want astronomy or mathematics, so later Run Run Shaw asked me to help him set up the Shaw Prize. We decided that these two should be given because they were two of the more accurate fields of science, so we gave them the awards. We added biology, because although biology already has a Nobel Prize, there are so many different kinds of biology now, and the impact of biology on human beings may be a little more direct than physics and chemistry, so we also gave one to that field. So, there are three Shaw Prizes awarded each year.

Ji Lizhen: Why do you think there is a Peace Prize? Why does the Nobel Prize have a Peace Prize?

Yang Chen-Ning: They also set up a peace prize. What's going on with the Peace Prize?

Ji Lizhen: It's been there since the beginning.

Yang Chen-Ning: It started with those three (prizes), then a literature (prize) and a peace (prize) were added, five prizes in total, after many years.

Ji Lizhen: It was in the 1960s that the economics (prize) was established.

Yang Chen-Ning: This economics (prize), you have to examine carefully, is not a Nobel Prize. It has a different name because it's not given by the Nobel Foundation. I think it's paid for by the National Bank of Sweden.

Ji Lizhen: Well, the last time I went to the Nobel Museum, I went into the hall, and there were some stories about Nobel.

Yang Chen-Ning: Is it in Stockholm?

Ji Lizhen: Stockholm. You haven't been there, have you?

Yang Chen-Ning: No, I haven't.

Nobel and His Girlfriends

Ji Lizhen: I went to Stockholm once, in February, and it was cold. I went around the city, and this museum was over there, so I just wandered in. There was an introduction to the Nobel Prize in the lobby. I found a picture of you and Lee Tsung-Dao. You looked so young! Then I turned to the small room on the right, which was the introduction to Nobel's life. I was shocked to see it. What was the matter? As soon as I entered, there was a photo of two very beautiful girls, with a sentence written on it. I still can't figure out this kind of thing, and we Chinese would never do this. How could they write it? There was a picture of a very beautiful and charming girl, who was said to be Nobel's lover for 17 years. Later, because someone else got the girl pregnant, they then separated. I wondered at that time if this really was a museum commemorating Nobel, because Chinese people would not write this kind of thing. Then I saw a picture of another beautiful girl, who was his secretary. Then one day when I went to the library to borrow some biographies of famous people for my daughter, I came across a biography of Nobel, so I borrowed it. After reading it, I looked for all the Nobel biographies I could find.

It was very interesting. There were a few of Nobel's letters on display in the Nobel Museum, and one of them was written to his girlfriend at the time, saying he was tired and had had a hard trip. Later, when I found the book and read it, I realized that his life story was really sad. I think every year when the Nobel Prize is handed out in October, everyone talks about who won the Nobel Prize, but most people have no idea about the story behind Nobel. It seemed that the story of Nobel and his girlfriend was quite complicated. Nobel was working alone in Paris, and he needed the help of a secretary, so an Austrian secretary came to help. The girl had been working as a governess in a wealthy family and had fallen in love with the family master's son, but she was firmly opposed by his family. Later, the girl left sadly and happened to see Nobel's advertisement. She went to work as Nobel's assistant for a week, and Nobel immediately fell in love with her. The girl told him that she had never forgotten her boyfriend, and Nobel told her that she should not miss him, because he might already have other girlfriends now. Later, Nobel went on a business trip. When he came back, the female secretary was gone. What had happened? When Nobel was on his business trip, the secretary's boyfriend wrote her a letter saying that he still wanted to be with her, so the secretary went back and eloped with him. After that, Nobel kept in touch with her as he liked her very much. Then the secretary worked on some peace movement, and according to the book, Nobel set up a peace prize, hoping that she would win the Nobel Peace Prize, which she won in 1905. The relationship between Nobel and his next girlfriend was such that after the secretary left, he returned home and was very sadly, so he went to the outskirts of Vienna to relax. In a flower shop, there was a very beautiful girl. This girl had been raised by her stepmother, and her own mother was gone. The stepmother had two daughters of her own, but the stepmother thought that the girl was very beautiful and that she was obligated to educate her well. She hoped that the girl could one day marry a very rich man. Later, she met Nobel, and he became her lover. After that, her family kept asking Nobel for money, which Nobel obliged.

Yang Chen-Ning: Have you ever read a biography of Nobel?

Ji Lizhen: I've seen five or six. I've read them all.

Yang Chen-Ning: I didn't notice the stories you were telling.

Ji Lizhen: And then there was the story that Nobel's family, who were against his marriage to this girl because he was 17 or more years older than her, maybe even 20 years. He didn't like her very much either. He thought she was uneducated, but he couldn't get away from her emotionally, so he bought her a nice house and locked her up in it. Nobel was often away on business trips, and the girl was lonely, so she found a man, and they became lovers, and she became pregnant. After the pregnancy was discovered, the girl wrote to Nobel, saying that she was pregnant with someone else's child and needed money.

Yang Chen-Ning: Are these in the Nobel Museum?

Ji Lizhen: They are in the museum. There's nothing on the museum's website about the girlfriend. You can see it when you enter the museum. His girlfriend didn't do well either. Nobel wrote a lot of private letters when he was dating his girlfriend. Nobel told her to destroy the letters immediately after reading, but his girlfriend didn't. After Nobel died, a foundation was set up. His girlfriend told the director of the foundation that she had some of Nobel's letters. One of the people in the foundation, who was Nobel's assistant, felt that it was not good for Nobel to have such letters published, so he told Nobel's girlfriend that he would pay for the letters.

Yang Chen-Ning: Later, there was a lawsuit.

Ji Lizhen: Yes, it was when her family divided up the property. Nobel's girlfriend had a little different opinion. Nobel's girlfriend sold some of her letters and later said she still had some. Nobel was also very miserable. You can see this about him in the letter … He looked like this. After the letter was sold to the foundation, it was supposed to be kept confidential for 50 years. Then, shortly after 50 years, a man wrote a biography in which he selected a lot of letters between Nobel and his girlfriend. I wrote an article on it because I thought it was so interesting. On the one hand, maybe we should thank Nobel's girlfriend. If Nobel's girlfriend had married him peacefully and had a child with him, I think there would have been no Nobel Prizes.

Yang Chen-Ning: The story you told about Nobel is at least different from the one I know.

Ji Lizhen: Haven't you heard it?

Yang Chen-Ning: People don't know much about it. This can actually be written as an article or introduction. Another thing that I think is worth introducing is a book that was published in English. I don't remember the title of this book. It told the story of awarding prizes in several decades after the Nobel Foundation was founded, maybe the first 50 years, I think. During those decades, there was a chairman of the prize committee who was a very politically powerful man with a very strong personality, so he presided over the awarding of the Nobel Prizes for a long time.

Ji Lizhen: It made a big difference, didn't it?

Yang Chen-Ning: He didn't major in physics, and the book had a description of what happened among the Nobel laureates who were influenced by this man. I can go back and look it up. If I find out the title of the book, I can tell you, but I'm afraid I can't because maybe I saw this book in the United States 30 years ago. I probably even had this book in the U.S., but I don't know where it was. After you read it, you will know that this person had a very unreasonable influence on the selection of Nobel laureates. Didn't I tell you that a biography of Fermi came out recently?

Ji Lizhen: No, is there a new biography?

Yang Chen-Ning: There is a new biography of Fermi in English recently. It's called *The Pope of Physics*. I told the physics department at Tsinghua University that part of the book was very good, and we should quickly translate it into Chinese, which I think they will do. What I'm going to tell you now is that the book talks a little bit about Lise Meitner. Do you know this woman physicist?

Ji Lizhen: Lise Meitner? Austrian, I think?

Wu Chien-Shiung and the Nobel Prize

Yang Chen-Ning: Meitner and her nephew, Otto Frisch, were the first to realize that uranium broke apart under neutron bombardment. This splitting is now called fission, where one splits into two. It was theoretically done by a German chemist named Otto Hahn, who wrote to Meitner about it. Meitner and her nephew thought for a while and said that Otto Hahn's interpretation of his experiment was not correct, and it should be fission. If you ask me, I think she could have won the Nobel Prize, but she didn't, so it's in the book, but it's not very clear. It just said that there was a very famous physicist in Sweden at the time named Karl Manne Georg Siegbahn, who was probably against Meitner. I read this and I was very interested in it. I told a guy at Fudan that he should study it. Why? Because there was another female physicist who didn't win the Nobel Prize, and people thought it was incomprehensible. She was Wu Chien-Shiung. Wu Chien-Shiung did not win the Nobel Prize, which I think will be studied in the future. But I have a conjecture, and this conjecture is related to this Siegbahn.

Ji Lizhen: Really? You mean he didn't like her very much?

Yang Chen-Ning: Well, Siegbahn himself was a Nobel Prize winner, about 20 years older than me. When I won the Nobel Prize, in 1957, he was the chairman of the Nobel Prize selection committee, so I knew him a long time ago. His name was Manne Siegbahn, and his son's name was Kai M. Siegbahn, who was also a physicist. His son and Wu Chien-Shiung were in the same profession, but Siegbahn was more famous in his field than Wu Chien-Shiung at the time of Wu's research, but Siegbahn did not do the experiment. My guess is that maybe Siegbahn was not satisfied with Wu because they were in competition with each other.

Ji Lizhen: Because of his son, I guess.

Yang Chen-Ning: He was already very established and was a leader in this field. I have no evidence that Wu Chien-Shiung was a competitor with Kai Siegbahn. I have no evidence at all, but I think he must have regretted not doing it. It turned out that 20 years later Kai Siegbahn himself won the Nobel Prize, but in a different field. I'm afraid he's dead now, and Kai Siegbahn is probably about my age. You see, his father was the chairman of the Nobel Committee for Physics at that time, and if he didn't get along with Wu Chien-Shiung, it would be easy to understand. I always thought it was possible, but suddenly I found a new biography about Fermi, and there was a suspicion that Manne Siegbahn didn't like Lise Meitner winning the Nobel Prize, which I think is something I can study.

I also want to tell you that this Meitner is Wang Ganchang's teacher. Do you know Wang Ganchang? Wang Ganchang is an important scientist in making atomic bombs in China.

Ji Lizhen: Oh, Wang Ganchang, yes, one of your classmates.

Jewish Scientists in the West

Yang Chen-Ning: He (Wang Ganchang) is a person who has been very successful; he is Meitner's Ph.D. student. I think there may be another reason why Meitner didn't win the Nobel Prize. It is because she was Jewish. At that time, it was not unusual to say that some people in Sweden did not like Jews.

Ji Lizhen: Jews are very influential now.

Yang Chen-Ning: Well, the attitude toward Jews in the world, especially in the United States, has changed since I was a student. When I was a student, the American attitude toward Jews, and the academic attitude toward Jews, was different than it was in Europe. Europe, Germany and Russia didn't like Jews, and that's still true in Russia today. You know a friend of mine recently passed away, Faddeev, whom I know very well.

Ji Lizhen: Is he a Jew?

Yang Chen-Ning: He is not Jewish, but he knows that there is a great struggle between Jews and non-Jews in Russian mathematics and physics circles. He always feels that he is involved. What does that mean? He thinks he doesn't discriminate against Jews at all, but Jews think he doesn't take care of Jews enough, so you can feel from his words that Jews and non-Jews in Russia are not satisfied with each other, which affects people's impressions of him.

Ji Lizhen: Yeah, Jews used to have a hard time finding work. Even when André Weil came to America, he had a hard time finding work at first, and Nottle also had trouble finding work.

Yang Chen-Ning: André Weil was made an example, so André Weil didn't like America his whole life.

Ji Lizhen: Right.

Yang Chen-Ning: Maybe it has something to do with Dr. Chern's most important job. André Weil was very unhappy in the United States in 1943 and went to Lehigh University, but Dr. Chern had been on good terms with him, so he talked to Dr. Chern and gave him a document of more than 100 pages. After Dr. Chern read it, he wrote the most important article in his life.

The Most Important Mathematicians

Ji Lizhen: Let me ask you another question. I'm doing some research right now to write a book, like a complete picture of mathematics. You have just analyzed for me what the important works in the history of physics are. I want to get the whole picture as best I can with mathematics, too. From your point of view, what mathematical results are particularly important, or what mathematicians are particularly outstanding to you?

Yang Chen-Ning: Of course, there are many who are older. After Gauss...

Ji Lizhen: Yeah, there seem to be a lot of people after Gauss.

Yang Chen-Ning: For example, mathematicians admire Abel very much.

Ji Lizhen: Yeah, Abel.

Yang Chen-Ning: I don't understand Abel. As far as I know, Abel was the first to prove that Gauss's equation couldn't be solved by roots. My impression was that he was going to work on integers, on things where the inverse of the square root was a cubic polynomial, an integer that he worked on a lot. Then Jacobi reversed it and came up with the elliptic equation.

Ji Lizhen: Yeah, Abel reversed it, too.

Yang Chen-Ning: Did Abel reverse it too?

Ji Lizhen: Yeah, he reversed it.

Yang Chen-Ning: Why do people keep saying it was Jacobi?

Ji Lizhen: The inverse of the Jacobian matrix is another problem, but when Abel and Jacobi studied the case where the genus is equal to 1, both of them had studied the ellipse. Later, when the Jacobian pair was a bit larger, there was a problem with the inverse of the Jacobian matrix. At that time, Weierstrass was famous. Weierstrass went from being a high school teacher to a professor at the University of Berlin. The special situation was just because of doing this.

Yang Chen-Ning: Who first discovered elliptic functions?

Ji Lizhen: It should have been Gauss, who knew it very early.

Yang Chen-Ning: Did you say that Gauss had already known about it?

Ji Lizhen: He had it all in his diary. Gauss mentioned a little bit in the last chapter of one of his famous mathematics books that Abel saw this and went on to study elliptic integrals, so we have elliptic functions and things like that from Abel and Jacobi. The problem was that he died too soon.

Yang Chen-Ning: But what about Abel's later influence? If you ask me, I know more about Galois's influence, because of Galois's group, but I had the impression that Abel was on the same level as him.

Ji Lizhen: Yeah, Abel should because of the whole complex analysis …

Yang Chen-Ning: Wasn't complex analysis done by the Augustin-Louis Cauchy and his colleagues?

Ji Lizhen: Abel is more important. Cauchy is also very important.

Yang Chen-Ning: What did Cauchy do that was important for the complex analysis?

Ji Lizhen: He also did something important and it was the Cauchy–Riemann equation, the integral formula.

Yang Chen-Ning: The Cauchy–Riemann equation is just an equation with a derivative …

Ji Lizhen: Yes, they give the conditions under which a function is an analytic function.

Yang Chen-Ning: What are the conditions under which the analytic function at that point, those two equations, that is …?

Ji Lizhen: Cauchy and Riemann, Riemann's paper.

Yang Chen-Ning: Does it have anything to do with Abel?

Ji Lizhen: No, Abel did it earlier, because Abel pioneered elliptic integrals, and he developed a more general one, which is called Abel's integral. And then there was Riemann, and Riemann's famous work was on Abelian integrals.

Yang Chen-Ning: So, if you ask me, I sort of understand that Galois has a huge influence, I understand that Riemann has a huge influence, and I understand that Cauchy has a huge influence. Let's say you want to ask me what Weierstrass did.

Ji Lizhen: Analysis is rigorous.

Yang Chen-Ning: I don't know. If you ask me what Julius Wilhelm Richard Dedekind did, I know Dedekind's partition theory is good, but I don't know what else he did. As for what you say Lie did, I can tell you about his influence. I can't tell you what Minkowski did.

Ji Lizhen: Minkowski did the geometric number theory.

Yang Chen-Ning: Minkowski influenced Einstein.

Ji Lizhen: Yeah.

Yang Chen-Ning: He was a number theorist himself.

Ji Lizhen: Yeah, geometric number theory. He created a new field.

Yang Chen-Ning: I don't know what he did in number theory. I'll tell you another thing: my father was a number theorist.

Ji Lizhen: You did tell me that.

Yang Chen-Ning: So, he had books on number theory on his bookshelf. When I was in middle school, I went to his bookshelf to study, and I saw a book written by Landau. Landau's book is very strange, very special.

Ji Lizhen: Why?

Yang Chen-Ning: Because we know that mathematicians and physicists need to write an article or write a book, that is, Theorem 1, Theorem 2, Lemma 1, Lemma 2. This is very clear and not uncommon. And then equations 1, 2, 3, 4. But usually when a person writes, say, a math textbook, the first chapter has some Theorem 1 and Theorem 2 in it; the second chapter has Theorem 1 and Theorem 2 again, so that each passage is independent, and those formulas are numbered 1, 2, 3, 4, etc. But that's not the case with Landau. Landau kept the number of theorems in order throughout his book, so that his theorems were often numbered (say) 135, and even more remarkable were his formulas, the number of which reached several thousand.

Ji Lizhen: Really?

Yang Chen-Ning: So, I think he is a very rare person. He requires accuracy. I know a lot of stories about him. For example, I don't know if you've heard this story.

Ji Lizhen: What story?

Yang Chen-Ning: He went to a class, which usually had very few students. One time when he went to class, there were no students, but he still finished his class. When he finished, he came out and saw one of his colleagues. He smiled and said, "Now they will never catch up." Have you heard the story?

Ji Lizhen: No, I haven't. I've heard that he was strict.

Yang Chen-Ning: But if you ask me what important work this Landau did, I have no idea. Having said that, let's come back to an important question: why are modular functions related to number theory?

Ji Lizhen: There are a number of reasons. For example, there is an important issue in number theory, which is that any number can be expressed as the squares of several numbers. It is a classical problem.

Yang Chen-Ning: It's the sum of the squares.

Ji Lizhen: Yes, and then count this number, which is related to the coefficient of the modular function.

Yang Chen-Ning: Is this kind of problem called a Waring problem?

Ji Lizhen: That's right.

Yang Chen-Ning: You say the Waring's problem is related to modular functions?

Ji Lizhen: The Waring's problem is a little bit different. The Waring's problem has a higher number of powers. I'm just talking about squaring, which is related to the function. The modular function that we talked about just now takes into account the expansion of the field—that's one thing. On the other hand is the L function, which was defined by Dirichlet. Let's say you count a number, you put everything together, systematically put it together, and you build an L function. The L function starts out in modular form, and then there's the comparison … This is also a big part of the Langlands program.

Yang Chen-Ning: What you're saying is, basically, that there's a class of number theory problems that involves studying these number theories from other theories?

Ji Lizhen: Well, for example, André Weil's most famous conjecture is that, given a variable, there are several points in the finite field. That's André Weil's most famous conjecture, the André Weil conjecture.

Because it counts numbers, every prime number, when you change p, you get a whole series of numbers. The question is, is there any formality, any pattern to these numbers? That's a good representation to show whether the L function constructed satisfies any properties.

Yang Chen-Ning: Is there any truth to such a statement? That is to say, people like Gauss or Abel, whom you just mentioned, thought about this number system, so there is the idea of a field, and different fields can give rise to different number theories. Moving in that direction, that's why a lot of great mathematicians are successful because they like to move in that direction. Isn't it true that the Chinese can't do this stuff?

Ji Lizhen: Classical Chinese mathematics has algorithms, lacking conceptual aspects.

Yang Chen-Ning: Right, so does it mean that these kinds of ideas are so impractical that the Chinese don't pursue them?

Ji Lizhen: I think all the learning in China is more practical. For example, if you are well educated, you can become an official.

Yang Chen-Ning: So, is Hua Loo-Keng less practical?

Ji Lizhen: The mathematics he did later, optimization, optimal method, which are (practical). His original mathematics may not be very useful, but what he did later …

Hua Loo-Keng and Siegel Compete

Yang Chen-Ning: Was there a period of time when Hua Loo-Keng was competing with Siegel a little?

Ji Lizhen: No, I don't think they could compete.

Yang Chen-Ning: Wang Yuan has said that in 1943, both Hua

Loo-Keng and Chern Shiing-Shen were invited to the Institute for Advanced Study.

Ji Lizhen: Right.

Yang Chen-Ning: Chern Shiing-Shen went, Hua Loo-Keng didn't. Wang Yuan has an explanation: it was because at that time Hua Loo-Keng was working in the same direction as Siegel, and he had concerns about being involved in confusion regarding the attribution of credit. Have you ever heard of that?

Ji Lizhen: No, I haven't.

Yang Chen-Ning: Haven't you heard it?

Ji Lizhen: I don't think so, because they're on a completely different level.

Yang Chen-Ning: I think Lin Kailiang told me that.

Ji Lizhen: No, I'll give you several reasons for my objection because Siegel's direction is different from Hua Loo-Keng's. Siegel does celestial mechanics, dynamical systems and automorphic types. Hua Loo-Keng did only additive number theory, which is more classical. In addition, I think their appeals were not on the same level. Siegel was older, and generally the older will not compete with the more promising young people. I don't think so anyway. Take André Weil, for example, I've heard that he's very unkind to his peers, but he's kind to some young people, so there was a reason, but Siegel didn't have to compete with Hua because Hua had just come out.

André Weil's Politics at the Institute for Advanced Study at Princeton

Yang Chen-Ning: Did I ever tell you the story about André Weil being unhappy with me later on.

Ji Lizhen: Really? I don't know. Why is that?

Yang Chen-Ning: As soon as André Weil went to the Princeton Institute, he immediately came to me, which happened probably in 1960 or 1961. Because he and Dr. Chern were good friends, he heard about me from Dr. Chern and he came to see me. I remember I was so surprised because the first time we talked, he said that he had heard of me. At that time, there was an old gentleman at the Institute for Advanced Study of history who was from Europe and who collected Roman coins. André Weil said to me, let's gang up and get him out here. I thought it was very strange, as this had nothing to do with him, so he did not ask me to do this again, and he did not succeed in this matter. So, at that time, I kept a respectful distance from him, but I really admired him. I didn't discuss any math or physics with him, and he never came to talk to me. We had lunch with other physicists, but he wouldn't join us. Mathematicians and physicists didn't have lunch together. We all had lunch in the same cafeteria. For instance, Jean-Pierre Serre and Borel, they would have lunch together, but those of us in physics were always together.

Ji Lizhen: When I went there, there was a table, where the mathematicians were sitting together.

Yang Chen-Ning: Why was he unhappy with me later? It was because he wasn't happy with Oppenheimer, and he and Montgomery were the two people who were most opposed to Oppenheimer. They were against Oppenheimer for several reasons, I think. One reason is that they looked down on Oppenheimer. They thought Oppenheimer was trying to have some kind of political influence, and they thought we should just do academic work, so they looked down on it. Another reason is that they thought Oppenheimer didn't raise money and wasn't giving money to the institute. It was said that Oppenheimer had made it clear to the committee that when he was in charge he wouldn't raise money, and the committee said it was okay, as they had a lot of money. But by the 1950s, there wasn't much money, so they turned against him. The opposition then focused on one issue, Milnor. Milnor was a brilliant young professor at Princeton. André Weil advocated poaching Milnor from the university to the institute. Oppenheimer said no, because in the early years of the Institute for Advanced Study, there was an unwritten verbal agreement (with Princeton) that the institute would borrow the Fine Hall (the mathematics department building) and that the institute would maintain a good relationship with the university and not poach people. So, no. André Weil and the others made all sorts of arguments that the unwritten rules didn't have to be followed, or that time had passed, years had passed, and everything had changed. In short, there was a lot of debate, and those of us in physics, of course, were helping Oppenheimer, so there was a lot of bickering.

Ji Lizhen: Yes, so André Weil was a little upset with you.

Yang Chen-Ning: It was a big fight. I remember once I was speaking at a teachers' conference, and I just said I'm a person from an ancient culture that had a less absolute view of the world, and everything had to be taken care of in many ways. Because I came from this tradition, I couldn't vote for Milnor in this matter. André Weil was probably not happy, he must have thought, okay, you bring out your Chinese culture, I have my French culture. How do I know that? He didn't say that. Many years later, there was a young physicist at the University of Chicago, a theoretical physicist, who is still around today, named Peter Freund.[56] I knew him quite well. He was a professor of physics at the University of Chicago. He was probably 20 years younger than me, and he had come to America from Europe. He became very familiar with this André Weil in some way, and I think the reason is because André Weil later spent a lot of time in Paris, and Peter Freund probably also spent a lot of time in Paris, so he became familiar with André Weil. About 20 years ago, this Peter Freund wrote a book, just a bunch of essays, and one of the things he said was that André Weil was not happy with me. What he said was that André Weil told this Peter Freund that Yang Chen-Ning was acting out of his Chinese culture. Because André Weil knew something about Chinese culture and had some respect for it, he

---

[56]Peter Freund died in 2018.

didn't have the same attitude as the average young American, he was probably particularly unhappy that I brought out my Chinese culture.

Ji Lizhen: Yeah, André Weil was a little bit …

Yang Chen-Ning: This guy was not easy to deal with, so people stayed away from him. But I think that if we were to name two or three of the most important mathematicians of the second half of the 20th century, I am afraid he would be one of them.

Ji Lizhen: Yes, he was very important.

Conflict Between André Weil and Grothendieck

Yang Chen-Ning: What about his relationship with Grothendieck?

Ji Lizhen: Not good, from what I've been told, it's because you could not have two tigers on one mountain. André Weil was trying to prove his Riemann conjecture on a domain of functions, then he laid down the foundation of algebraic geometry, and he was pretty much the No. 1 of the world. Then along came Grothendieck, who knocked it all down.

Yang Chen-Ning: Do you mean Grothendieck went a little further?

Ji Lizhen: Yeah, a lot further, and now that André Weil's set is no longer used, it's said that Grothendieck came and knocked it all down.

Yang Chen-Ning: I heard that Grothendieck wrote a book and scolded a lot of people.

Ji Lizhen: Yeah, it's a thousand pages, and we plan to translate it.

Yang Chen-Ning: Can you give me the title of the book?

Ji Lizhen: Yes, it's in French.

Wang Liping: It's called *Harvest and Sowing* in Chinese.

Yang Chen-Ning: You say it has been translated into Chinese?

Wang Liping: No, some people have translated it and put it on the Internet.

Yang Chen-Ning: Can you send it to me?

Wang Liping: Yes, I can send you the link.

Ji Lizhen: Did you know Serre is coming to Tsinghua next month?

Yang Chen-Ning: Serre, I know a little about him, but I'm not familiar with him.

Ji Lizhen: Serre and Grothendieck had a lot more contacts, and Serre asked him why he scolded his students. Because he had a student, Pierre Deligne, whom he liked very much.

Yang Chen-Ning: What's his name?

Ji Lizhen: Pierre Deligne, now at Princeton.

Yang Chen-Ning: I don't know him.

Ji Lizhen: This is a young man. He's Grothendieck's best student. What Grothendieck had always wanted to solve was André Weil's conjecture, to develop a theory, and his student made it without using his theory. He felt betrayed by his students, like Jesus and Judas. They quarreled violently. Serre said they both had an explanation: they both had strong personalities. About André Weil and Grothendieck, I also asked a few French people at the time. André Weil was the most important person in the Bourbaki school, and when André Weil gave the Bourbaki lecture, Grothendieck didn't answer his questions very well. Later, when Grothendieck gave the lecture, André Weil was probably not very friendly in his questioning, so Grothendieck couldn't stand it, and he went out and left the Bourbaki lecture. The two men were not convinced by each other, so it was an example of not allowing two tigers staying the same mountain. Neither could tolerate the other. Grothendieck also left mathematics very early, because he was a very strange man. He scolded a lot of people in his books, but the people who were scolded felt very honorable. My teacher's teacher was also a professor at Princeton. His name is Robert MacPherson. You don't know him. He is younger.

Yang Chen-Ning: What's his name?

Ji Lizhen: Robert MacPherson.

Yang Chen-Ning: Oh, Robert MacPherson is still there?

Being Scolded by Grothendieck is Not Something Discouraging but Honorable

Ji Lizhen: Yeah, he's (MacPherson's) there now. He went there in the 1990s. At the end of Grothendieck's book there was a group of people carrying a coffin to bury Grothendieck, and my teacher's teacher and two colleagues from my department were among the people carrying the coffin. They were both very happy about it, just to be able to say that he only scolded those who were qualified. It was a very strange relationship.

Yang Chen-Ning: The tension between mathematicians is worse than in physics.

Ji Lizhen: Really?

Fewer Disputes Among Physicists Than Among Mathematicians

Yang Chen-Ning: The reason is that there are many details that can be left alone in mathematics, so some people are willing to leave this, while others are willing to leave others, and there are many quarrels. There is no such thing in physics. Physics is either right, valuable, or worthless. It is not easy to argue about it.

Ji Lizhen: That's right, and then I'll give you an example. You know, there are quite a lot of conflicts among Chinese mathematicians in China, and people abroad also know that. Andrew John Wiles was famous for solving Fermat's theorem. One of my colleagues at that time was a student of Wiles. He said, it's not only you Chinese who have conflicts in the family, we have them too. He said, look at our big family, Wiles's teacher is Coatcs, Coates's teacher is Baker, Baker is the Fields Medal winner, Baker and Coates have a bad relationship, Coates and Wiles have a bad relationship, and Wiles has a bad relationship with his student Taylor. Four generations: Baker, Coates, Wiles, Taylor. Every other generation is very good. I think in human nature, it seems that every other generation doesn't matter. So, when you said that Siegel competed with Hua just now, I don't think the reason is very convincing.

Yang Chen-Ning: I think the relationship among mathematicians is not the same as that among physicists.

Ji Lizhen: And I have a question for you. I think physicists are more intelligent, that is to say, they have a better understanding of worldly things. Many mathematicians seem to be from another planet.

Yang Chen-Ning: The reason for the difference is that what is good in mathematics that can be left behind, is completely different in nature from that of physics. What mathematics can leave behind is a very beautiful and complex structure, and this structure is growing. But physics is not like that. Physics fades very fast; it's like the desert. If it becomes sand, there would be nothing to fight about.

Ji Lizhen: Less to quarrel over.

Yang Chen-Ning: Actually, this can be a topic for an article. It's very interesting. Do you like the articles written by Cai Tianxin and Tang Tao? They want to make a magazine, a bit like *American Mathematics Monthly*, also a bit like *Mathematical Intelligencer*. It is somewhere between the two, and I think this deserves encouragement. But their articles are not as good as those in *Mathematical Intelligencer*.

Ji Lizhen: Dr. Yang, I have a question here about Grothendieck. Grothendieck left the IHES (Institut des Hautes Études Scientifiques) at the age of 42, and then stopped doing his previous mathematics. What was the reason? His official reason was that he found out the IHES was using military money, which he felt was immoral, and he left. How did Serre say that Grothendieck was tired of doing math on his own? Because it actually took him so much time to do it. He was going to build a big, big, cathedral-like structure, but it took him decades only to build a wall footing. What do you think of this? Does it also relate to whether a person can be productive all the time? Because he's still young at 42. After he left, he scolded his students for not going through with his whole plan. May I ask, what is your opinion on this matter? Is there such a person in physics who leaves while he is at the top of his career?

Yang Chen-Ning: I agree that there are such kind of people, but few go to such extremes as whom in mathematics you mentioned. For example, Grothendieck and Perelman, as well as Stephen Smale, are somewhat like this.

Ji Lizhen: Smale looks okay, I think.

Yang Chen-Ning: The U.S. government didn't like Smale very much.

Ji Lizhen: Yeah, because he was against the Vietnam War.

The Unusual Smale

Yang Chen-Ning: Many mathematicians admire Smale very much.

Ji Lizhen: Really? Why?

Yang Chen-Ning: I don't know why. Many years ago, some mathematicians said that Smale was amazing.

Ji Lizhen: Smale was a strange man. I once invited him to the University of Michigan to join a colloquium. Several students were scheduled to talk with him. He said that when he was a graduate student at Michigan, he almost failed an exam, and the dean came to him and said if he didn't do well next time, he would have to leave there. The reason, he said, was that he had a good friend at the time, and the two of them had a contest to see who could do worse without getting kicked out of the department. It was like riding a bike, going a little slower than the other guy. When he won the Fields Medal, he made some remarks in Moscow against the Soviet government. I asked if he was afraid at that time, he said he was not afraid because he thought he had the Fields Medal, and anyway, they didn't dare to do anything to him. So, I thought he seemed interesting.

Yang Chen-Ning: Is he going to Hong Kong SAR now?

Ji Lizhen: Yes.

Yang Chen-Ning: Going now?

Ji Lizhen: Yeah, the last time I ran into him was at the Riemann Conference three years ago. He was in good health. We had dinner on the beach in Sanya the other day. I said we could go for a walk on the beach. Later, we came out of the restaurant. There was a quite high fence. We were ready to climb over. I asked if he could, and he said it was no problem. He could do it, too, and then he followed us over.

Yang Chen-Ning: Chinese cultural tradition discourages and dislikes people who go to extremes. If you ask me, I think there are advantages and disadvantages. What are the advantages? I think the Chinese tradition has something to do with China's rapid development over the last few decades after the Reform and Opening-up because (it can't be that) too many people are developing in all directions. China is going from a situation of extreme poverty to a situation where all need to work together to strive to move in one direction. It is not advantageous to use the Western way. But is there a downside to encouraging China's way too much? There is. So, it's very complicated. Just like the ongoing debate about whether the Chinese education system is better than the American education system, the average Chinese people think the United States is better, while the American people think China is better.

Ji Lizhen: Dr. Yang, would you mind if I send you some questions and you write some responses?

Yang Chen-Ning: You can email it to me or give it to me.

Ji Lizhen: Yes, I have already given it to you.

Yang Chen-Ning: If I can comment, then I'll comment.

In 1985, Chern Shiing-shen received an honorary degree from SUNY at Stony Brook. From left: Zhang Shoulian, Chern Shiing-shen, Yang Chen-Ning

Ji Lizhen: Okay, if you can comment, I will send this to you because some of the questions I have raised may be more detailed, so please write down some of your thoughts about them if you can.

Yang Chen-Ning: Okay.

Ji Lizhen: Okay, thank you. We'll talk to you next time.

The Fourth Interview

Time: July 4, 2017
Place: Institute for Advanced Study at Tsinghua University
Interviewer: Ji Lizhen, Wang Liping
Recorder: Wang Liping
Collator: Peng Cheng, Wang Liping, Ji Lizhen

Institute for Advanced Study, Tsinghua University

Everyone knows that symmetry is important in life, and beauty is often associated with symmetry. So, what's the math behind symmetry? How did this happen? Group theory, or rather transformation group theory, can be used to understand crystal structures. Yang Chen-Ning and Lee Tsung-Dao won the Nobel Prize in Physics for their theory of parity non-conservation, which is still symmetric. In this interview, you can gain insight into Dr. Yang's deep understanding of symmetry, its history and its application to physics.

Mathematics Is Created Alongside but Independent of the Universe

Yang Chen-Ning: I think mathematics is independent of the other phenomena in nature. In other words, since the Big Bang, there has been mathematics in seconds, and there has been mathematics in microseconds. In other words, mathematics is an independent structure. Why do you have this phenomenon? I don't know how to answer that. Mathematics itself has symmetry in it. For example, in two dimensions there is no difference between left and right, but in three dimensions there is a difference between left and right. Do you know that Weyl wrote an article? It's a popular article, probably a preface to one of his books, and he talks about the distinction between left and right as one of the most important and fundamental phenomena. So, I think there is symmetry between the physical world and the biological world because the structure of the physical world and the biological world is based on mathematics. Since mathematics has an idea of symmetry, nature has an idea of symmetry.

Ji Lizhen: As you said just now, the establishment and development of mathematics is independent of nature.

Yang Chen-Ning: For example, if we find people on other stars, there must be mathematics too. Their mathematics may not be the same as the mathematics language here, but the structure is the same. I don't believe it will be different.

Number of Crystal Classes

Ji Lizhen: The other thing you said about the application of mathematical symmetry in physics is that Emmy Noether worked out a conservation law because the Lie group was acting on it, and then a conservation law emerged.

Yang Chen-Ning: My view of symmetry in physics research goes something like this. The first major breakthrough is crystal symmetry.

This was quite natural, for, unlike mathematicians, physicists derived their knowledge of the solution of symmetry, and of group theory, entirely from phenomena. Mathematicians derived their knowledge from the structure of mathematics; physicists didn't pay special attention to the structure of mathematics. The most natural thing in physics was crystal symmetry, and to study crystal symmetry, physicists learned that there were many different families. They went back and forth, and finally, around 1890, a Russian physicist named Vladimir V. Fedorov, and two mathematicians, Arthur Moritz Schönflies and Auguste Bravais, had a breakthrough. Firstly, they got the idea that there were crystal families. Then, from a mathematical point of view, it was found that there were crystal families from the group structure. As physicists discovered different families from the properties of crystals, they again reached the same destination. By 1890, there was a long debate about how many families there were, for there were more than 200 of them, some of which were not immediately clear as to whether they were the same.

Ji Lizhen: Yeah, about 270, I remember.

Yang Chen-Ning: I only remember that it is more than 200. You know, there is a famous story like this, after understanding the three-dimensional space—I don't remember who came up with the idea, someone questioned why not ask the same question in two-dimensional space. There were 17 of them in two dimensions, so this idea of two dimensions came about after the study of three dimensions. I wrote an article on this. Let me think. I'm afraid it's this one.

Ji Lizhen: Really, the two-dimensional conclusion came later? It's a beautiful picture.

Yang Chen-Ning: The physicist was Vladimir V. Fedorov, and then the mathematicians were Arthur Moritz Schönflies and Auguste Bravais, and they figured out that every crystal family was related to a space group, and there were about 230 different space groups in three dimensions. So, they got 230, and then the 17 in two dimensions. Next came Hilbert's 23 problems, one of which was whether there is a finite number of crystal families in any dimensional space.

Ji Lizhen: Yeah, finite. Ludwig Bieberbach solved that later.

Marie Curie and Symmetry

Yang Chen-Ning: Bieberbach solved this problem later. But how many four dimensions are there? It takes a computer to figure it out. As far as I am concerned, no specific number has yet been obtained for five or six dimensions, only upper and lower bounds. There are many people who specialize in studying such things. This is the general history. In addition, there is another important thing you said about symmetry today that is worth studying; you know Pierre Curie, the husband of Madame Curie? He wrote a lot of articles about symmetry. He was a rather clever man, but I don't think it made any difference. If you want to study this history, you have to look at what he was up to.

Ji Lizhen: I didn't think of that. It seems that because Madame Curie is more famous, her husband was neglected, wasn't he? That's how I feel.

Yang Chen-Ning: I remember that the most important work in his life was about symmetry, but it had almost no effect. I think he mostly studied symmetry from a philosophical perspective.

Ji Lizhen: That's interesting. I should take a look.

Einstein and Symmetry

Yang Chen-Ning: The concept of true symmetry has had an impact in two different directions. One direction was by Einstein in 1905.

Ji Lizhen: Lorentz's?

Yang Chen-Ning: After he wrote it, a year or two later, Minkowski said that his Lorentz transformation symmetry was a four-dimensional,

3+1-dimensional group. Einstein didn't like that. Einstein wrote an article saying that what Minkowski was doing was superficial, but two years later, he completely changed it.

Ji Lizhen: Yeah, and then he spoke very highly of Minkowski.

Yang Chen-Ning: No, he felt like … In his later years, he talked how he started working on general relativity, and I have an article here on Einstein that tells the story. On page 272, "Opportunity and Perceptions" I quoted from an article Einstein wrote in his later years. He said that it was only a year or two after Minkowski that he realized that this symmetry was important, and that he wanted to write down the dynamical equations. These equations were transformation invariant, which led to general relativity. So, you could say that symmetry made a fundamental contribution to the building blocks of physics, and that this first shot was made by Einstein. But ordinary physicist didn't really appreciate it, and the experimental physicist didn't really feel much of an impact from it. From the perspective of experimental physicists, group theory began to become important with the atomic spectrum. After the discovery of quantum mechanics, experimental physicists found that the atomic spectrum had quantum numbers, so did spectral physicists.

Ji Lizhen: Yes, it corresponds to a group representation.

Application of Weyl, Wigner, and Lie Groups' Representation Theory to Physics

Yang Chen-Ning: Actually, physicists do not have the concept of this group; they only have the concept of quantum numbers. What is a quantum number? A spectral line, for example, divides into three when placed in a magnetic field, so the spectral line is related to three. Another spectral line in a magnetic field becomes two, so this is related to two. Therefore, many quantum numbers are obtained from the spectrum, and the beginning of quantum numbers, I think, was at the end of the 19th century, or you could say that the Germans were doing spectra very accurately, and they got a lot of numbers. It took decades to work on these numbers, which were very mysterious. But by the end of the 1920s, two most important books came out, one by Weyl and the other by Wigner, saying that these numbers, the quantum numbers of the physicists and the experimental physicists, corresponded to some natural numbers of group theory. I think from a historical point of view, this finding really began to put symmetry as a concept at the center of physics.

Ji Lizhen: Einstein was the first.

Yang Chen-Ning: Einstein's didn't affect experimental physicists; he affected pure (physics) theorists. From the standpoint of experimental physics, the most important thing is what I just said. It turns out that the number of quantum numbers is closely related to the group.

Ji Lizhen: I think mathematically, there are two differences. Lorentz groups are noncompact, and elementary particles are compact Lie groups, right?

Yang Chen-Ning: This distinction works for mathematicians, but doesn't work for physicists. But what I didn't say just now is that Wigner's and Weyl's books were both very famous, but ordinary physicists, not only experimental physicists but also theoretical physicists, didn't really understand them. And it's not just ignorance, it's resistance. So you know there's a term called gruppenpest. You've heard of it. A lot of people, especially physicists, didn't get into group theory at the moment, but they knew there's something wonderful about it, so they didn't like it very much. For example, one of the most important gruppenpest (theoretical approach to abstract groups) was discovered by John Clark Slater. Slater was a pretty good theoretical physicist, but he was more traditional. He didn't come from mathematics. He later discovered the matrix Slater determinant, and he wanted to make a wave function that was antisymmetric, and writing it as a determinant was, of course, antisymmetric. He suddenly found that he could use matrices to understand a lot of phenomena instead of group theory, so he sort of said in a letter that he had now overcome the gruppenpest. Here's a wonderful book, a classic book in the 1930s. You could also say it was a book on the relationship between group theory and atomic physics, so it became a monograph. When I was a graduate student, everyone read this book. This book explained the very complex phenomena of atomic physics from the point of view of a physicist by means of mathematical group theory.

Ji Lizhen: Then people changed their view.

Yang Chen-Ning: This book is different from Weyl's and Wigner's. Their books are more mathematical. Weyl is more mathematical, and Wigner is more physical. His book is for people who read physics.

If you want to study history, you should read the book *Atomic Physics and Group Theory*. Do you know this journal? It's called *Reviews of Modern Physics*, and it's still being published.

Ji Lizhen: This one's from 1948, and there's another one from 1936. Okay, that's it.

Yang Chen-Ning: Yes, this article deals with molecular physics and group theory.

So, for fundamental physicists, both theoretical physicists and experimental physicists, these two are the most important. It tells you how atomic physics gets started, namely which is how group theory and atomic physics get started, and they are molecular physics and group theory.

I studied group theory and physics as an undergraduate, and when I was writing my undergraduate thesis, and I was writing my bachelor's thesis at Southwest Associated University, I went to Wu Ta-You. As I mentioned in this book, he asked me to read this book, read this article, so I read. Now my paper is in the archive (https://arxiv.org), and I'm still surprised to read it, because it was written by a fourth grader.

Ji Lizhen: It's amazing that a college student could write this kind of thing.

Yang Chen-Ning: I understood this long article and summarized the mathematical structure of the article. By the end of the 1930s, everyone understood the importance of group theory. So, in other words, usually, people who study physics, especially experimental physics, understood neither Weyl nor Wigner, but they all started with these two books.

In the 1940s, and after the war, they built on it. One of them was Giulio Racah. Do you know the Racah coefficient? Also called 6j-symbol.

Ji Lizhen: Is it some kind of coefficient from group representation theory?

Yang Chen-Ning: And 9j-symbol.

Giulio Racah was a Jew and an Italian, and when there was a war, he escaped to a place near Israel. There, he studied groups, and one of the most important things in groups was SU(2), or SO(3), which had irreducible representation. He multiplied the two irreducible representations, decomposed them, and he got the Clebsch–Gordan coefficient. Do you know the term?

Ji Lizhen: Yeah, I've heard of it.

Yang Chen-Ning: The Clebsch–Gordan coefficient. Regarding this kind of problem, in this book, Racah studied hard.

Ji Lizhen: How does this work in physics? I've never thought about it.

Yang Chen-Ning: If an electron, its state, its …

Ji Lizhen: Corresponding to a group representation?

Groups Representation and Elementary Particles

Yang Chen-Ning: Since its Hamiltonian conforms to SO(3) symmetry, its eigenvalues are to belong to an irreducible representation. Therefore, different orbitals belong to different irreducible representations. Now add two electrons, one belonging to an irreducible representation, the other belonging to another irreducible representation. It's an orbital precession that belongs to a 2×2 representation, and the other one belongs to a 4×4 representation, and when you put these two together, you're multiplying them; you're multiplying a 2-dimensional representation with a 4-dimensional representation. The multiplication can be divided, and that's how you get the Clebsch–Gordan coefficient. It means that you need to figure out how to multiply two different irreducible representations.

Ji Lizhen: Yeah, my question is, what does it mean physically? Mathematically, I know. Physically, are two particles colliding?

Yang Chen-Ning: An electron, its spin is ½, which is irreducible representation, namely 2×2. Another electronic spin is three halves, whose orbit is three halves, namely 4×4. When these two electrons encounter, there's an interaction.

Ji Lizhen: What happens after the interaction?

Yang Chen-Ning: After they interact, they break down. If you ask about this two-electron system, you break these two up, so you have to calculate this thing all day long.

Ji Lizhen: After the two interact, it's all about the decomposition. I see.

Yang Chen-Ning: So, from a mathematical point of view, two irreducible representations are multiplied and decomposed. The coefficients of the decomposition of many irreducible representations are called Clebsch–Gordan coefficients. This Clebsch–Gordan coefficient has a lot of mathematical properties, so Racah worked on it, and that's how it came out; it's called 6j (-symbol), 9j-symbol, and then it came out in a book. I told you it was an important book, or an important article, and then Racah published a third important book, called *Elements*. It's a little book. If you read the book, you'll know about 6j-symbol, and 9j-symbol. The 6j-symbol is two irreducible representations multiplied by each other, and the 9j-symbol is three irreducible representations multiplied by each other, and then of course there's the 12j (-symbol). Some people worked on that, so Racah became very famous later. *Elements* was not so famous, but his book was very famous because it's the kind of book that theoretical physicists—especially theorists who do this sort of thing—read all day.

Ji Lizhen: Does it have something to do with Yang–Mills?

Yang Chen-Ning: Not really. The Yang–Mills theory is related to Einstein.

What this does takes place at the ground level and has something to do with the experimental medium. Gauge field theory asks for general principles, so it's a different type.

Ji Lizhen: Different type. Is it group theory?

Yang Chen-Ning: They surely were all connected later. But from a historical point of view, they were two different rivers. These books would be very interesting if you take a look.

Ji Lizhen: Yeah, I will because I've heard about it mathematically, but I've never thought about how it works physically. It happens when two particles collide.

Yang Chen-Ning: In fact, for any group, there are Clebsch–Gordan coefficients. There must be somebody in this field doing it. For any group, say SO(3), there are many irreducible representations, and when you multiply these irreducible representations, it also has Clebsch–Gordan coefficients. The other one related to this is an Italian who later became a professor at the Institute for Advanced Study called Stephen Adler, in physics, who is still around and younger than me. He found an equation in this Racah coefficient, which I suspect is the same for any group [I don't know] whether he generalized it or not.

Ji Lizhen: So, this is group representation theory, and compact Lie groups are finite groups, which are very useful in the elementary particles of physics. What you've just told us is that the representation of compact Lie groups is very important.

Yang Chen-Ning: In fact, the direction of physics is not the representation of compact Lie groups, just SO(3). We're especially interested in SO(3).

Ji Lizhen: It seems that this has something to do with the quark environment, right?

Yang Chen-Ning: Yes, after the war ended, particle physics started to develop.

The war lasted until 1945, just during this generation, and then a new field came out called particle physics, and I happened to be a graduate student at that time, and as soon as particle physics came out, new fundamental particles were discovered. You've probably heard of it. It's called a strange particle.

At the initial point, there were one or two strange particles. Then more were found, so Oppenheimer later said that strange particles were a zoo. He called them a zoo. The question was how these strange particles interacted with each other. There would be symmetry in this interaction, so we studied the symmetry of particle interaction. This was the most important problem in the 1950s. We used a lot of group theory. All we talked about was three-dimensional rotations, which we already knew. At that point, it was discovered that there needed to be new symmetries. Why you have new symmetries? Because there was a spectrum. These elementary particles had symmetry, and since there was a spectrum, it must be an irreducible representation of some group, so we used irreducible representations of some group here. First we got SU(2), then SU(3), and we went on. These were things done in the 1950s, and I worked on these things. But I asked a question that others didn't ask at the time, which was whether they developed basic principles.

Ji Lizhen: To explain so many things.

Lie Groups and Challenges Facing the Yang–Mills Theory at Its Initial Pitch

Yang Chen-Ning: You know that there should be some groups, but when you know groups, how can you use groups to determine interactions? It's electrodynamics. You can write this Maxwell equation. Now, if you want to turn it into SU(2), what equation should you write? So my question is whether the structure of the equation is SU(2) or SU(3), so that's the Yang–Mills theory.

Ji Lizhen: Yeah, because you mentioned it last time in a recollective article, and in a video. Within ten years since you made this result, it seemed that no one had realized its importance.

Yang Chen-Ning: The reason was that there was a problem immediately after it was made. It seemed that the theory was going to create a particle which had no mass. It's not a photon, which had no mass. I still remember when I gave a speech to people, they would ask me, "Why can't we find this very important massless particle?" So, I couldn't answer this question.

Ji Lizhen: I see. That's the main question.

Yang Chen-Ning: But in hindsight, I thought I had to write out both this and Mills. I also published it in *Physics Review* at that time, and they accepted it without asking any questions.

Ji Lizhen: Really? Is it because it's beautifully written?

Yang Chen-Ning: I don't know why, but I don't remember anyone inviting me to give a talk on this before the 1970s.

Ji Lizhen: Yeah, I saw in the video that you gave a talk at Harvard University, and you chose the topic yourself.

Yang Chen-Ning: That's because Harvard University was looking for a new assistant professor, so they asked people of my age one by one to lecture there on topics of their own choice. I gave this lecture, and nobody asked me any questions after that.

The Age of Harvest for Physicists and Mathematicians

Ji Lizhen: So how old were you when you worked on the Yang–Mills theory?

Yang Chen-Ning: 32 years old.

Ji Lizhen: 32 years old, that's not easy, because 32 years old is old for some people.

Yang Chen-Ning: I went abroad late because of the war. I actually graduated from college when I was 20 years old, but I was delayed, so I got my doctorate when I was 26 years old.

Ji Lizhen: Great job. I was reading this book this morning. Schrödinger's great work was done at the age of 39. He said it's relatively rare in physics, although it did happen.

Yang Chen-Ning: I think mathematicians need to be younger than theoretical physicists. Like Galois and Abel, they were all very young. I'm afraid Selberg was very young, too.

Ji Lizhen: Yeah, Selberg was very young when he started working on "An Elementary Proof of the Prime Number Theorem."

Yang Chen-Ning: I think he was 30 years old.

Ji Lizhen: Yeah, 30 years old. Serre was relatively young (he won the Fields Medal at the age of 28). There was no one particularly young behind him.

Yang Chen-Ning: In 2006, I took my wife, Weng Fan, to the Institute for Advanced Study to show her the place. I found Selberg in the common room of the institute, sitting there reading a newspaper. He didn't recognize me.

Ji Lizhen: Really? It has been too long.

Yang Chen-Ning: Actually, I had worked with him for more than ten years, and my wife had known him. I asked him, "How is your wife?" He said his wife had passed away. She was also Norwegian and had come with him from Norway to America, so I knew her better. He told me that he got married again. I hadn't met his new wife. We took some pictures, but I can't remember if it was a picture of me and him sitting next to each other or just him. He was gone the next year.

Ji Lizhen: Your work on gauge field theory is great. Does your work have something to do with quarks and stuff like that?

The Relationship Between Resolution of Challenges Facing the Yang–Mills Theory at Its Initial Pitch and Renormalization Group

Yang Chen-Ning: There were some follow-ups. Well, after Mills and I wrote this article in 1954, because the problem of massless particles had not been solved … And then in the 1970s, someone came up with a new idea, and this new idea was called symmetry breaking. This idea of symmetry breaking was proposed in the 1960s, and by 1970, it was found to be correct. Did you know that gauge field theory had a renormalization problem? Renormalization had actually been around since the 1930s, and quantum electrodynamics had basically been written up by the 1930s. But if you do the mathematics, it diverges because in quantum mechanics, you have an electronic state of something, and it's affected by other electronic states. There are an infinite number of other electronic states, and you have to add up the infinite number of effects. In the 1930s, however, it was calculated in this way, and the results were infinite, so it was called the infinite obstacle. This obstacle was discovered around 1947. It is infinite, but it is possible to limit these infinite quantities to a finite number. This is called renormalization. As for the most important contribution, I think it's Dyson, but he didn't get the Nobel Prize, which I think is very unfair. The reason is that those three people won the Nobel Prize, but none of them had enough mathematical influence, and Dyson had mathematical influence, so it was Dyson who deserved it. All three of them started working on this, but they didn't have the whole problem figured out, which Dyson had the ability to do it.

Ji Lizhen: I heard renormalization is needed when the Feynman integral is infinite. Is that it?

Yang Chen-Ning: Yeah, Feynman, Schwinger and Tomonaga Shinichirō, they all started to do renormalization, but they could only do the lowest ones; the higher ones they couldn't figure out, so they didn't do them. Dyson learned how to do it from Feynman, and then he used Feynman's algorithm to figure out that higher-order ones diverge. But there was an idea, which was not Dyson's. This idea was called renormalization, and it was Dyson's job to carry out this renormalization. So, by 1950, between 1951 and 1952, quantum electrodynamic renormalization was more or less understood, and that meant it was actually an unfinished job from the perspective of mathematicians.

Ji Lizhen: Right, it was not strictly proved.

Yang Chen-Ning: Because the whole thing was not very definable. In a Hilbert space with infinite degrees of freedom, the math hadn't been figured out yet, but physicists had a way to figure it out, and it's amazing. When it came to Dyson's quantum electrodynamics renormalization, people started with the second order, then the third order, and the fourth order. Now physicists have probably reached the fifth order. The calculated result is one billionth of the accuracy of the experiment.

In other words, you found a very smart graduate student, you taught him QED (quantum electrodynamics) and renormalization, that is, to calculate the fourth order. If he was very smart and would spend several months on calculating it, the calculated results would be consistent with the tenth to ninth order.

This meant that QED was right, and renormalization was reasonable. By the 1950s, although it was not so accurate, everyone knew that renormalization was an important thing.

And then gauge field theory came and needed to be renormalized, and I had to renormalize when I did it, although I didn't know how, because it was very complicated, and a lot of people did it. I did it, and Feynman did it. Finally, he did it first by invoking the Feynman diagram of non-Abelian gauge field theory, and this one was done by Faddeev.

Ji Lizhen: Oh, Faddeev. You said he died not long ago. Did he do this renormalization?

Yang Chen-Ning: Faddeev and his student Victor Nikolaevich Popov wrote an article, and then Hoft used their method to prove that the theory of non-Abelian gauge field was renormalized. This was very important (work), because people believed in renormalization, which made people pay more attention to the theory of non-Abelian gauge fields. So, two things happened in 1970. One thing was that Hoft showed the theory of non-Abelian gauge fields was renormalizable, which was a purely theoretical idea. And the other thing was that the experiment was done, and it said that SU(2) × U(1), the non-Abelian theory, was consistent with the experiment. So, by the beginning of 1970, when these two things were discovered together, people were convinced that the non-Abelian gauge field theory was more than just a framework. SU(2) × U(1), this non-Abelian gauge field theory, moved from the theoretical to the experimental. Through the above two developments, everyone accepted this, and there was experimental proof, which happened in the 1980s.

Ji Lizhen: So how did this relate to symmetry breaking?

Yang Chen-Ning: Using this and adding this.

Ji Lizhen: You meant this symmetry breaking plus …

Yang Chen-Ning: But not this one. There was a question of no mass. To solve the problem of no mass dispersion, a symmetry breaking was added artificially to get a mass. So, in my point of view, this was a very dissatisfying thing.

Ji Lizhen: Yeah, what does symmetry breaking mean? You changed the group?

Symmetry Breaking

Yang Chen-Ning: No, symmetry breaking is not like that. Symmetry breaking means the structure is still symmetric, but when you look at it from an asymmetric direction, you find something strange. Hence, symmetry breaking …

Ji Lizhen: So, it's symmetric, and if you look at it from an asymmetric direction, it's not very symmetric.

Yang Chen-Ning: Like a round head—if you look at it sideways, it's not symmetrical.

Ji Lizhen: Yeah, when you look at it from this side, it looks like a circle, but when you look at it sideways, it looks like an ellipse. That's what it means.

Yang Chen-Ning: So, the meaning of symmetry breaking is actually a symmetrical thing, but you change your point of view, then it can have mass. But this method is very … It's an artificial way to look at it, and that can't be permanent. It has been successful so far, but there must still be … This, of course, is the one-million-dollar prize, which is to ask what to do about it.

Ji Lizhen: How did you solve it?

Yang Chen-Ning: Change the question. In other words, how did mass come about? Symmetry breaking is very successful now. The experiment has been successful, but no one thinks this is the last thought.

What you just asked was that when you got here, you added the two together and explained QED and weak interactions. QED plus weak interaction was explained, but strong interaction was not. Strong interaction was to generalize this asymptotic freedom. And that triggered something that's not very clear. That didn't fit very well with the experiment, which is now called the Standard Model.

Ji Lizhen: The Standard Model. That's what David Jonathan Gross and the people after him did, didn't they?

Yang Chen-Ning: Theirs is SU(3).

Ji Lizhen: The Standard Model is right here. In fact, I've heard that all the unusual Lie groups in physics are coming in now.

Yang Chen-Ning: So, a lot of people studied it at that time, and there were a lot of articles on it, but none of them have had anything to do with experiments so far.

Ji Lizhen: What are they going to do? Are they going to unify all of them?

Yang Chen-Ning: Not necessarily. Well, some people want a grand unified theory, and some people want more than that.

Ji Lizhen: Yeah, and gravity.

Yang Chen-Ning: This has not been successful so far, so you can imagine that with some results, of course, inevitably, people go to ...

Ji Lizhen: So, the gauge field theory is very important. You see, here it's just a different group. Anyway, the main theory here is gauge field theory, so this is a different group—U(1)SU(2) here, times another one here, has basically the same framework, right?

Yang Chen-Ning: Yeah, it's just that the group gets bigger and bigger, and why it is like this? There is no explanation.

Standard Model and Noether's Work in Symmetry

Ji Lizhen: That's interesting. What you were saying, the Standard Model of particle physics, that's completely different from Noether's theory, or is it related?

Yang Chen-Ning: There is a close relationship.

Ji Lizhen: Really, how?

Yang Chen-Ning: Well, Noether's work was in the 1910s, and physicists weren't paying attention at that time. But when these things came along, physicists knew that group theory was important. Then there was a problem. Physicists had known about conservation before Newton, but it was Noether who pointed out the relationship between conservation laws and symmetry. Noether's relationship was between symmetry and conservation laws, and by this time people knew that symmetry was important, so it was even more important for Noether. I think it influenced some ideas, but it didn't really affect any concrete results. So, by the time I went to graduate school, everyone knew that symmetry was closely related to the number of quantum particles and the law of conservation. Take parity, for example.

Ji Lizhen: Yes, parity is a violation …

Yang Chen-Ning and Wigner

Yang Chen-Ning: Well, here's what happened to parity. How did astronautics discover it? It was because the people who were doing atomic spectroscopy suddenly found a professor at the University of Michigan named Otto Laporte. He was an experimental physicist, I think he was a theorist, not doing experiments himself. He studied the experimental results of atomic spectroscopy, and then he discovered something called Laporte rule. Laporte rule says that there are two states of an atom, one called odd and one even. So, the transition must go from odd to even, or from even to odd; you can't go from even to even, and you can't go from odd to odd, and that's called Laporte rule. Laporte rule was proved by Wigner. "In 1927 Wigner took the critical and profound steps to prove that the empirical rule of Laporte is a consequence of the reflecting invariants of electromagnetic forces." This is a very important statement. Why? Because, you know, before Wigner said this, many physicists looked down on him, thinking that he was too mathematical and had nothing to do with physics. He was unhappy at Princeton for a while, but he was actually quite a deep thinker, so he left in a fit of rage and went off to Wisconsin. Then it was years before he was invited back. In the 1960s and 1970s, he said that when I was young, no one thought my work was important. Today, everything I do is considered important. This was in my Nobel Prize speech, and I made it very clear. He was very happy because it was the first time that I mentioned the idea of parity was proposed by him. I won the Nobel Prize in 1957, and a few months later Princeton gave me an honorary degree, which was proposed by Wigner.

Ji Lizhen: Well, Wigner felt that your work was a continuation of his own.

Yang Chen-Ning: I think I did the right thing, because I think Wigner's problem was that experimental physicists didn't appreciate his stuff; they thought it was too mathematical, and mathematicians thought everything was done by Weyl, and he didn't have much, so he wasn't as important as he was in the 1930s. But when you look at this, the idea of parity is different from other symmetries because classical approaches didn't have that idea, where a symmetry produces some quantum numbers because it's continuous. It's a discrete symmetry, it also has a quantum number, and this is parity. I think he's very proud of his contribution, and he deserves to be proud of it.

Ji Lizhen: Did Wigner win the Nobel Prize for this work?

Yang Chen-Ning: No. It's very interesting that he won the Nobel Prize later. Look it up. He won the Nobel Prize without a good reason.

Wigner was a very honest man. Let me give you an example. You know that the field theories are used now by physicists, but before 1930, some field numbers used in field theory were operators, but these operators were real numbers. Now it can be anticommutative, not just ordinary numbers. Now these variables of field theory can be anticommutative, and this anticommutative mathematical system was developed by Jordan and Wigner.

Jordan Algebra Named After a Physicist

Ji Lizhen: Yeah, Jordan Algebra, that's him, the physicist. When I first heard about Jordan algebra, I thought he was a mathematician, but then they said no, he was a physicist.

Yang Chen-Ning: Jordan made two important contributions. One was after Heisenberg's one-man paper, Jordan and Born wrote a two-man paper. Then Heisenberg came back and they wrote a three-man paper. So, quantum mechanics began with these three papers.

Ji Lizhen: Really, Jordan made a great contribution.

Yang Chen-Ning: Yes, one of Jordan's important contributions was using anticommutative variables with Wigner, which Fermi later quoted in field theory in the 1930s. Mathematicians are not like physicists. Physicists are against this kind of thing, but when everyone else has done it, the physicists stop thinking about it and accept it again. Physicists are less thoughtful than mathematicians. However, you can see from now on that it is a very important contribution that Jordan and Wigner made by using anticommutative variables. I remember, I have an article on this. I'll show you later.

Ji Lizhen: Okay, thank you.

The Fifth Interview

Time: July 30, 2018
Place: Institute for Advanced Study at Tsinghua University
Interviewer: Ji Lizhen, Wang Liping
Recorder: Wang Liping
Collator: Peng Cheng, Wang Liping, Ji Lizhen

Science is created by human beings. Learning about the life of a scientist deepens our understanding of science. Whether in China or abroad, the interpersonal relationships of scientists have a great impact on scientific research and can make the world of science more complex or interesting. The academic dispute between Chern Shiing-Shen and Hua Loo-Keng may be less discussed in China, but it does exist. There was also a lot of unpleasantness between Weyl and Hua Loo-Keng. Writing a biography is itself a science. In this interview, Dr. Yang shared how his biography was written and the difficulties he encountered. Dr. Yang also commented on the biography of Hua Loo-Keng written by Wang Yuan. Dr. Yang mentioned his quarrel with Lee Tsung-Dao, and Lee Tsung-Dao's influence on the physics department at Columbia University.

Ji Lizhen: Dr. Yang, the photo I took last time at Yunnan Normal University was of your quote "The real secret of success is interest." I think this sentence is particularly useful in the real world because I feel that now this society has become a little too pragmatic, and many people don't want to learn basic subjects like mathematics and physics anymore. They are more interested in making money in business.

Yang Chen-Ning: This is especially true for Chinese students because there is a philosophy behind the whole social education system in China, and this philosophy is discipline. Discipline means that it is imposed on you. But outside China, especially in the United States, more attention is paid to stimulating interest, which is a very complicated issue. The American system has its advantages and disadvantages, which needs to be studied in depth.

Ji Lizhen: Yes, recently, science in the United States is still quite advanced because people there have many ideas. The day before yesterday I went to CITIC Bookstore, a new bookstore that just opened in front of Tsinghua University, I was surprised to see that 90 percent of the books were all translated. I would like to ask why we Chinese don't write our own books? All depend on translating books from overseas? I think this is not a very good phenomenon. We have so many people. We should write books by ourselves. I think it has something to do with interest. Americans really like …

Yang Chen-Ning: It's necessary to develop a tradition of writing books for ourselves. For example, writing biographies, or writing biographies of scientists. China doesn't have this tradition, while the West likes to write biographies. China has had a tradition of biographies since Sima Qian. This tradition has its merits, but in the present era of scientific and technological development, it is not good because it is not specific enough and does not provide a clear analysis. Here I would like to start with a story. I think one of the best biographies of this kind in Chinese was written by Zhang Dianzhou. Do you know Zhang Dianzhou?

Ji Lizhen: Yes, he wrote about you, and he wrote about Chern Shiing-Shen.

Yang Chen-Ning: He is the professor at East China Normal University.

Writing Process of Wu Chien-Shiung's Biography

Yang Chen-Ning: I'm sorry, it's not Zhang Dianzhou, it's Jiang Caijian. Zhang Dianzhou also did some very interesting work, but he didn't write biographies. I haven't read his biographies. His work is also important in this regard, but today I'm going to talk about Jiang Caijian. Jiang Caijian is now, I think, 80 years old. He used to be a reporter for the *China Times* in Chinese Taiwan, and worked as a war correspondent in the Middle East. The owner of the *China Times* at that time was Yu Jizhong, who is no longer with us. 余纪忠 in Chinese. He was the owner of the *China Times*, one of the two largest newspapers in Chinese Taiwan. He was a classmate of Wu Chien-Shiung at Central University in the 1930s, so he thought he should write a biography of Wu Chien-Shiung.

In the late 1980s, he approached the journalist Jiang Caijian and said, "I will pay to send you to the United States, where you will live near Columbia University to find Wu Chien-Shiung and her husband and then write a biography of her." So Jiang Caijian went to America. How did he go about writing it in America? He went to see me in Stony Brook first, because he had visited me in Taipei before. When he got to Stony Brook, he told me what he was going to do, and I told him I thought it was worth doing. I told him I would take him to a bookstore to buy some biographies written by Westerners. After reading them, he would know that this method of writing biographies is worth learning. So, I took him to the bookstore.

Ji Lizhen: Which ones did you buy? Do you remember?

Yang Chen-Ning: There might be a biography of Hilbert, all in English. And sure enough, he wrote her biography. It took him several years to write his Chinese biography of Wu Chien-Shiung, which should be translated into English.

In fact, there was a famous physicist named Rabi, whose grandson studied Chinese, who knew Jiang Caijian, so he said he would do it.

But he had not written it down for many years. Why Rabi's grandson didn't write it, I don't know; it's a pity. After the biography came out, Jiang Caijian came to me and said he would write one for me, and I agreed. So, this biography of mine was written in 2002, and it's very well written. What's special about it is that he interviewed over 100 people. We still have his tapes, and a lot of the people on that tape were gone. For example, my classmate at the University of Chicago named Marvin L. Goldberger, who later became president of Caltech, was interviewed, and that recording is still there. On this recording, you'll see that every quote was annotated, showing the date and the place.

Ji Lizhen: That's very good. It's of historical value.

Comment on Hua Loo-Keng's Biography

Yang Chen-Ning: The two biographies written by Jiang Caijian should be translated into English. No one has ever done this, but it is very necessary. I think the books need a good translation in English. Writers are not quite aware that this kind of biography written in the Western way has its advantages. For example, *The Biography of Hua Loo-Keng* by Wang Yuan is great, but it only has one biography, and it is not detailed enough. As for Hua Loo-Keng, if Hua Loo-Keng were an American, there would, no doubt, be several biographies, thick biographies. In China until now, among messy biographies, I think the one standing out is the biography written by Wang Yuan. But Wang Yuan is not a writer. He is a mathematician, and he only knows some of the information.

Ji Lizhen: Yeah, I don't think his descriptions of the characters are vivid enough. There are some other things, too. He's more mathematical.

Yang Chen-Ning: That's right. In fact, I have advocated for this idea before, but nothing came of it. I had some contact with several foundations, and I once suggested that they set up a fund specifically to encourage young people to develop in this direction. Thus, this could make them stop looking around, and in a few years, settle down to do it. But none of this has any real effect.

Wang Liping: Our publishing house has found a foreign professor to translate Dr. Hua's *Advanced Mathematics*.

Sensitive Issues on the Simplified Chinese Version of *The Biography of Yang Chen-Ning*

Yang Chen-Ning: After *The Biography of Yang Chen-Ning* was published in Chinese Taiwan, in 2002, the publisher approached several publishers on the Chinese mainland, including Tsinghua University Press, to publish a simplified Chinese version, but nothing came out of it.

The reason was that there was a dispute between Lee Tsung-Dao and me in it. The official stance is to avoid involvement and not to bring it up here, so it didn't come out. Ten years later, in 2011, I thought it was ready to be published, so a simplified Chinese version was published. However, if you compare the simplified Chinese version with the traditional Chinese version, you will find that some content has been deleted. I did not check carefully; I only noticed that one paragraph was deleted. Why? It had something to do with politics. You know there was an important Chinese physicist named Wang Ganchang, who's dead and was one of the 23 most important people in China's atomic bomb experiments. In the book Jiang Caijian wrote about me, and a passage about Xu Liangying and Wang Ganchang. Do you know who Xu Liangying was? Xu Liangying was a philosopher and a writer. He was the one who translated the complete works of Einstein. It was probably in the 1990s that Xu Liangying, influenced by some anti-Chinese writers in the West, went to Wang Ganchang. He was Wang Ganchang's student, so Wang Ganchang had a good impression of him. In the meantime, he said some things in foreign newspapers that Wang did not say. After reading this, I was not satisfied, so I went to investigate the matter and wrote a letter to Wang Ganchang. Later, Wang Ganchang wrote back. The whole paragraph was about me, Xu Liangying and Wang Ganchang, and about some false rumor in the West ...

Money and Publishing

Yang Chen-Ning: I would like to ask your publishing house this question: do you lose money or make money?

Wang Liping: The publishing house makes money.

Yang Chen-Ning: Because you publish a lot of higher education textbooks?

Wang Liping: Yes, we publish textbooks for higher education.

Yang Chen-Ning: For example, *Introduction to Calculus* or something like that. Do you make money by this?

Wang Liping: We do, but lose money on publishing academic books.

Yang Chen-Ning: Because you can make a lot of money here, are there many curious phenonmena?

Wang Liping: What curious phenomena do you mean?

Yang Chen-Ning: This is the case not only in China, but also in the United States. How do I know? We talked about a famous physicist named Feynman, who said in one of his books that he lived in California, and somehow got involved with the press in California. He learned that in order to make money, the California press was doing something that was clearly not in accordance with the ethics of education, and Feynman blew the whistle on it. Feynman liked to do that kind of thing. So, wherever there is money, curious things happen, both in China and abroad, not just in China.

History of Southwest Associated University

Ji Lizhen: I'm very interested in Southwest Associated University. I looked for some books. Have you seen the one by John W. Israel?

Lianda: A Chinese University in War and Revolution. Have you read this book?

Yang Chen-Ning: I have. In fact, he worked on it for many years. He started writing it in the 1980s, so he came to visit me in Stony Brook, and then probably some problems occured, and it took many years to complete. I think it was more than 20 years before it was published. After I read it, I was a little disappointed. I thought that he didn't know enough because there were a lot of rumors in the book [somewhat different from] the real situation. What is the real phenomenon? The real phenomenon that Yang Chen-Ning saw can't be the same as that of a person who studies literature. You know there is a book called *Weiyang Song*. Do you know about that one?

Ji Lizhen: I know about it, but I haven't read it carefully. It's from Chinese Taiwan.

Yang Chen-Ning: *Weiyang Song* was written by one of my classmates who studied literature. I hardly knew him, but I know there was such a person. I think his book represents part of the life of my classmates at that time.

Ji Lizhen: It's fairly realistic, isn't it?

Yang Chen-Ning: At Southwest Associated University, there were talented people in all fields. Some people wrote books about the atmosphere of Southwest Associated University at that time, why Southwest Associated University was successful, what was special about Southwest Associated University, and what was not successful about the university.

For example, there was a series of books written by one person. He had collected a lot of materials in the past few years, but he didn't do much sorting or analysis. He just put the original materials together. He went to interview some people, called … I don't remember, but there were several books.

Nankai University Compared to Swarthmore College

Ji Lizhen: Let's check out one book. I noticed one weird thing. The writer said, suppose Cuba takes over Washington in the United States, and there are three universities in the United States to be consolidated. One is Harvard, the second is Yale, and the third is Swarthmore. I think Swarthmore is equivalent to Nankai University, but people from Nankai University might wonder at this match because Harvard and Yale are famous, and Swarthmore is only known to a few people in the United States. What does the writer mean by this analogy? I wonder why he compares Nankai University to Swarthmore.

Yang Chen-Ning: I didn't pay attention to that. Now that you are talking about it, it is not appropriate because the characteristics of the three schools in China at that time were completely different from those in the United States, so the three universities had different features. And because of these different characteristics, there is some friction, but there are also some successful aspects of the relationship, which are completely different from those (dynamics) in the West.

Ji Lizhen: Yeah, I wondered why did he give that example of these three American universities?

Yang Chen-Ning: I think the example is from an American educator. All in all, I thought he could write a good book at the beginning, but I was very disappointed after reading it. A very important point about these three universities, which I think is not mentioned by anyone, is that both Peking University and Nankai University had an advantage over Tsinghua University. Both of their presidents were in high positions in China's educational circles, but Mei Yiqi did not have a special position. However, Tsinghua's specialty was that it had money and, therefore, equipments. In fact, it had already made some preparations to ship some things to Changsha around 1935 and 1936. Later, many instruments came from Tsinghua. I thought in advance that the north might be unstable, so later, I kept in touch with the Tsinghua Research Institute at Southwest Associated University. At that time, there was a Tsinghua University Research Institute in Kunming, in the suburbs near where I lived. There were very important scientists doing various types of research in biology, electric machinery and physics. The reason they were able to do this was that some equipment had been shipped in. All of this was rooted in funding for Tsinghua University, not from the Ministry of Education, but from the Ministry of Foreign Affairs. So, when I lived in Tsinghua Garden in the 1930s, Peking University and Normal University, especially Normal University, were often unable to pay salaries. Tsinghua did not have this problem. That's because Tsinghua's funds came from the Boxer indemnity. The Boxer indemnity came from the Ministry of Foreign Affairs, and how did the ministry get the money? Because in the Boxer indemnity agreement, there was a clause stating that China's indemnity was paid by installments. How to ensure that the installment payments come in place every year? So, there was a clause actually worth checking. I remember that one of the reparations was because of the Shanghai Customs Office, the general director of which was a British man, who later became very famous. The man took control of a large part of China's financial power in the era of the Republic of China. According to the highest financial priority of the reparation, indemnity money should first be sent to the Boxer Indemnity Committee, which ended up at Tsinghua University. This man is so famous that there is a bronze statue of him on the Bund in Shanghai. I remember this statue was only removed when the Japanese occupied Shanghai or after liberation in 1949. This event is poignant in Chinese (historical) stories. This man wrote a letter which basically meant that no Chinese and no dogs were allowed in the public garden on the Bund. However, few people have written about this man's bronze statue and why there is one.

Ji Lizhen: Is it because he worked for China?

Yang Chen-Ning: I think you need to find out who sent him the letter of appointment. I don't know.

Hua Loo-Keng and the Doctorate Dissertation of Yang Chen-Ning's Father

Ji Lizhen: Actually, I pay more attention to the mathematicians of Southwest Associated University. One is Hua Loo-Keng and the other is Chern Shiing-Shen. The day before yesterday, I read a little biography of Hua Loo-Keng. It was stated in the encyclopedia that Hua Loo-Keng came to Tsinghua University, and it was your father who guided him to learn number theory. Later, Hua Loo-Keng began to work on number theory.

Yang Chen-Ning: Pardon?

Ji Lizhen: Just after Hua Loo-Keng came to Tsinghua, it was your father who guided him to learn number theory, and then Hua began to do number theory.

Yang Chen-Ning: The first articles he published were about …

Ji Lizhen: Waring's problem.

Yang Chen-Ning: Waring's problem. His earliest articles are all at Tohoku University. If you go to check the *Tohoku Mathematical Journal*, you can find his first few articles, and they're all about Waring's problem. This is because my father's paper is about Waring's problem. Do you know what my father's paper is about? Was it Gauss who proved that any positive integer was the sum of four squares?

Ji Lizhen: Lagrange proved it first.

Yang Chen-Ning: Yes, you're right, Lagrange proved it, and then it was said that every positive integer was the sum of nine cubic numbers. It might be Leonard Eugene Dickson who proved it. After Dickson proved it, he told my father that it could be proved that any positive integer was the sum of nine pyramidal numbers. The pyramidal number is a pyramid of a pile, so the smallest one of a pyramid is one, the bottom one is a four, the top one of the bottom three balls is a four, then the bottom one is a ten, which is one, three, six, and this number is called the pyramidal number r. The number of the n-th pyramidal number is $1/6(n^3 + 3n^2 + 2n)$. So, it's the cube number. But it was more than the number of cubes, so Dickson asked my father to prove it, and my father proved it. Then I wrote an article about this history. Some guy after him proved it. It seemed that my dad proved it was all nine, but the other guy, after many years, proved it only needed eight, and then seven cubes. And then, in the 1980s or the 1990s, I went to study my father's work, but nobody understood because they had his paper at the University of Chicago, and his paper was so long that I thought it would take me weeks to figure out how to prove it, so I didn't. But I was able to use a computer to verify it, and I ended up writing a paper with a prediction that it would only take five, or maybe only six, and then after a very high number, it would only take four. The collaborator with whom I wrote this article was a computer scientist. He did some more calculations with a big computer, which proved that our prediction was correct and that we could get to a larger number.

Later, Enrico Bombieri came to Stony Brook, and I told him about it. I asked him if there was any way we could take this one step further. He thought about it and said no. With new methods, problems like this are not particularly easy to solve.

Ji Lizhen: That's nice. I went to the museum of Southwest Associated University in 2012. I've been there twice. The old campus is Southwest Associated University, and recently, I went to the new one. When I went there last time, I went specifically to see if there was anything in the museum that introduced Hua Loo-Keng's first prize in science. I wanted to take some photos.

Competition and Conflict Between Chern Shiing-Shen and Hua Loo-Keng in Southwest Associated University

Ji Lizhen: I heard that there were some conflicts between Hua and Chern in Southwest United University because of that award, which affected the development of the whole Chinese mathematics field.

I am curious about Southwest Associated University. What do you know about Hua Loo-Keng and Chern Shiing-Shen? You are around the same time as them.

Yang Chen-Ning: I can't say that I know much about Hua Loo-Keng and Chern Shiing-Shen. I can only say that I know more about some specific parts.

Between them personally, I think it is like that. First, they are competitors, and second, they are two very different people. And that, I think, has a lot to do with their family backgrounds. I don't think it's fair to say that there was a conflict between them in Kunming or afterward, not even in Hua and Chern's private lives, but there was a fierce rivalry between them, which was obvious. And there was intense competition before Chern Shiing-Shen went to Princeton in 1943 because the general impression was that Hua's work was better than Chern's.

Ji Lizhen: Really? That's why Chern didn't get any science prizes afterwards?

Conflict Between Hua Loo-Keng and Weyl

Yang Chen-Ning: There are a lot of (possible) answers because at that time, no Chinese scientists understood his work, and his work in the United States, especially with André Weil, was immediately known to be very important. China was not at this level, so no one in China understood it then. Hua was the type of person who wanted to sell himself very strongly, but Chern was smarter than him. Of course, everyone wanted to spread their name, but Chern was smarter. He didn't need to go about it in a stupid way ... I don't understand what Hua has done to offend Weyl. Every time Weyl saw me or saw my wife at a party, he would criticize Hua Loo-Keng, so it must be that after Hua Loo-Keng left, I guess Weyl felt perhaps Hua Loo-Keng used his ideas and did not give credit to Weyl for his suggestion.

Ji Lizhen: He didn't give credit to Weyl.

Yang Chen-Ning: Because Weyl was already the first person, the leader, in that modern age. How could you do that?

Ji Lizhen: Hua Loo-Keng was quite young at that time. Because he was a self-made man, he had to fight for it all by himself.

Lee Tsung-Dao and the Department of Physics, Columbia University

Yang Chen-Ning: There is now a consensus that Lee Tsung-Dao was solely in charge when he was at Columbia. The university was very successful in the 1950s because of Isidor Isaac Rabi, and there were a lot of Nobel Prize winners at the university, some of them were winners then, and some of them won later, such as Gurdev Khush, Willis Eugene Lamb, Rabi and so many others. As a result, by the 1960s, everyone had left for a reason, and the only Nobel Prize winner who stayed was Lee Tsung-Dao, so he became very powerful at Columbia. At that time, Columbia University was full of stories, and people who went there to give lectures knew about them in advance, so they had to be very careful because Lee Tsung-Dao would ask a lot of questions. At that time, there were several things that Lee Tsung-Dao might have been wrong about. One was Steven Weinberg and the other was Ding Zhaozhong. Neither of them could stay at Columbia University. The story was written in Jiang Caijian's book. Jiang Caijian said in the book that he went to visit Rabi's wife after Rabi was no longer alive, and Rabi's wife said that the Columbia University physics department was founded by Rabi, but it was destroyed by Lee Tsung-Dao. There was this (specific) statement in the book.

Ji Lizhen: The physics department at Columbia University doesn't seem very good right now.

Yang Chen-Ning: It's not bad, but it's not as good as Stanford. There is a Nobel Prize winner there, a Chinese, who grew up in Henan, went to school in Hong Kong, and then went to the United States to do experiments in condensed matter physics.

Ji Lizhen: He's still young?

Yang Chen-Ning: No, I think he's 80 years old. If you look up the Nobel Prize winners, and you look up the Nobel Prize winners around 2000 (Daniel Chee Tsui, Cui Qi, 1998 Nobel Prize winner) (you can find him).

Ji Lizhen: It is not easy to maintain a department that has been established. You said that Hua Loo-Keng's (story) had something to do with the background he grew up in. Was Hua Loo-Keng self-taught?

Yang Chen-Ning: Yes, and you can say that after he did some of Waring's problems with my father, he went on to analytic number theory, and then he went to Hardy. I don't think anybody in China knew anything about analytic number theory at that time. He was self-taught, so when he came to Cambridge University, Hardy admired him very much. He was a genius, and that's no doubt about this.

Influence of the Debate Between Chern Shiing-Shen and Hua Loo-Keng on Chinese Mathematics

Ji Lizhen: I don't know about this situation. I read in a book that the competition between the two men later influenced some things in Chinese mathematics.

Yang Chen-Ning: The competition between Chern and Hua certainly had an impact on the later development of Chinese mathematics, but it was not very important, because there was also this (same) kind of competition between foreigners.

Ji Lizhen: I heard that Chern suggested closing the math institute after he returned to China.

Yang Chen-Ning: I don't know. I'm not sure if that is true or not. Can you find out about it in any document?

Ji Lizhen: I don't know. This sounds surprising to me because the math institute was already doing so well.

Yang Chen-Ning: Based on what I know about Chern Shiing-Shen, he wouldn't say anything like that. Does he think there is something bad about the math (institute)? I don't know much about it, and it's not impossible for me to (imagine) it … What I do know is that Chern did not come back to Tsinghua.

Why Chern Shiing-Shen's Institute of Mathematics Is Not in Beijing

Ji Lizhen: Yes, I'm curious. Why did he go to Nankai?

Yang Chen-Ning: Why Nankai?

Ji Lizhen: Yeah, I don't think Nankai was convenient.

Yang Chen-Ning: Yes, it's not just inconvenient. He knew that the Chinese academic center was not in Nankai. Why didn't he come to Beijing? We talked to him about it, but I could sense that he felt that by coming to Beijing the competition with Hua would be obvious, and he felt that it would not be easy … That's my guess. I think it goes without saying.

Ji Lizhen: Do you think Chern just didn't want to compete, or did Hua have a big influence at Tsinghua?

Yang Chen-Ning: Not just at Tsinghua, but all over Beijing. Chern Shiing-Shen happened to graduate from Nankai, so he went back to Nankai. From what Chern Shiing-Shen said to me, I got the impression that he thought Hua had decided to return to China in the early 1950s because Hua suddenly felt that Chern's scientific status in the United States had become very high, and he wanted to develop in another way. Chern Shiing-Shen didn't tell me explicitly, but I felt that when he was talking about himself and Hua's life, including Hua's returning back to the country at that time. I felt that he thought it was the reason. Can it be just I feel this way? No, because after all, I'm not a math major, so I think it is the impression Chern left on me. He felt that although he did not make it clear, there were certainly many reasons why Hua resolutely returned to China at that time. Was it at that time when Chern was invited to give the one-hour lecture?

Ji Lizhen: Yes, this happened.

Yang Chen-Ning: He didn't feel like competing with Chern in the U.S. anymore. You can ask Wang Yuan. Is Wang Yuan all right now?

Ji Lizhen: Yes, he is. I just met him the day before yesterday. He's editing the *Encyclopedia of Chinese Science*. He's in charge of the math volume.

Yang Chen-Ning: He's living in Changping, to the north of Beijing. Isn't that a nursing home?

Ji Lizhen: No, I met him at the math institute. He isn't living in a nursing home during that time.

Yang Chen-Ning: You said he doesn't live in a nursing home?

Ji Lizhen: I don't think he lives in a nursing home. I met him at the math institute. I met him on the road.

Wang Liping: It was said so before. I always thought he was in a nursing home, but Mr. Ji said he met him.

Yang Chen-Ning: I read an interview in which some of his students went to visit him in a group. He was very proud of himself. He said that before he moved into the nursing home, he had done a lot of research and thought it was good, so he moved in. Haven't you seen this article? Cai Tianxin said it was published in their magazine.

Wang Liping: *Mathematical Culture?*

Yang Chen-Ning: Is there a group of magazines like *Mathematical Culture* in Jiaotong University in Chinese Taiwan?

Wang Liping: *Mathematics, Science, History and Culture.*

Yang Chen-Ning: It's a serial publication. You're not the editor?

Ji Lizhen: I'm not an editor.

Yang Chen-Ning: You know Liu Zhaoxuan was a good friend of Ma Ying-jeou at that time. More than good friends, they had a very deep relationship, and Liu Zhaoxuan was a very hands-on person. So, as soon as Ma Ying-jeou became the leader of Chinese Taiwan, he immediately appointed Liu Zhaoxuan as his deputy, and that is Chinese Taiwan's number two position. It happened that the academic conference was held in Chinese Taiwan that year. Ma Ying-jeou had just become the leader of Chinese Taiwan and Liu Zhaoxuan had just become the deputy, so I still remember that. When the meeting was held, Liu invited several foreign academicians to dinner, including me and probably Shen Yuen-Ron and other individuals. At that time, Liu Zhaoxuan and I both thought that there would be some important achievements, so Liu Zhaoxuan was very ambitious at that time. But you know, there was a flood. A few months later, there was a flood in Chinese Taiwan and probably some people died, so he lost his official position.

Ji Lizhen: Really? Because he didn't handle it well?

Yang Chen-Ning: He only did it for several months.

Ji Lizhen: Really? People blamed him?

Yang Chen-Ning: He still has some influence on the Nationalist Party in Chinese Taiwan. I think this is also an unfortunate thing in his life. He is a very capable and intelligent man, and he is a good communicator.

Ji Lizhen: And his Chinese was pretty good. He often writes articles about mathematics and poetry.

Yang Chen-Ning: His kung fu novels are the best in Chinese Taiwan.

Ji Lizhen: He writes kung fu novels?

Yang Chen-Ning: He's writing again now.

Ji Lizhen: Really? I didn't even know that. I know he used to have a lot of stories about math and poetry. Because he was at the Institute of Theoretical Physics, they started the Institute of Theory in Chinese Taiwan, didn't they?

Yang Chen-Ning: Do you have a close relationship with Chinese Taiwan?

Ji Lizhen: Not so close, but sometimes when I was writing a book I edited (I used to write a book about mathematics and humanities), then Professor Liu also gave us a manuscript. I do have several classmates in Chinese Taiwan, but nothing else.

Yang Chen-Ning: Are there any successful students from Chinese Taiwan who have gone to the United States to study math these past years?

Ji Lizhen: Yes, H. T. Yau. Have you heard of him?

Yang Chen-Ning: The one at Harvard?

Ji Lizhen: Yeah, he's doing probability and studying the conjectures of Dyson.

Yang Chen-Ning: I met him once because one of his articles was related to one of mine, about statistical mechanics.

Ji Lizhen: Statistical mechanics? Probably he works on random motion.

Yang Chen-Ning: Where did he get his Ph.D.?

Ji Lizhen: Princeton.

Yang Chen-Ning: You said he went there from Chinese Taiwan?

Ji Lizhen: First to Princeton, next to the Courant Institute for postdoc, and then to Stanford, because he knew Hsiang Wu-Chung pretty well. You only met him once. I think he came from Chinese Taiwan and now performs best.

Yang Chen-Ning: What about Liu?

Ji Lizhen: Liu what? Liu Tai-ping? My feeling is that Liu Tai-ping did a good job with PDE (partial differential equations), but he didn't seem to have as much impact H. T. Yau have. Actually, I have a lot of questions about Southwest Associated University. Do you know Xu Lizhi?

Southwest Associated University and Xu Lizhi (Hsu Leetsch C.)

Yang Chen-Ning: Of course.

Ji Lizhen: Xu Lizhi (Hsu Leetsch C.)[57] has written a lot about Southwest Associated University, and he wrote a book about Hua Loo-Keng.

Yang Chen-Ning: Xu Lizhi (Tsui Lap-Chee) of HKU, you say?

Ji Lizhen: No, from Dalian University of Technology. Xu Lizhi (Hsu Leetsch C.), just think about it. He's 97 years old now.

Yang Chen-Ning: Hsu Leetsch C.? He can write very well.

____

[57]Hsu Leetsch C. (1920–2019) was a Chinese mathematician and educator.

Ji Lizhen: He is very good at writing, and he told a lot of stories about Southwest Associated University and Hua Loo-Keng.

Yang Chen-Ning: He really knows about Southwest Associated University.

Ji Lizhen: Yes, people told me that it seemed that he wrote a book about Hua Loo-Keng, but Hua Loo-Keng's family was not satisfied with it.

Yang Chen-Ning: Is he still alive?

Ji Lizhen: Yes, I met him last year at Tsinghua University. I met him last year at the 90th anniversary of Tsinghua University. I talked to him.

Yang Chen-Ning: I met him in Kunming.

Ji Lizhen: Really? You knew him in Kunming? Of all the people I know, you two are both at a ripe old age and still in good spirits. Do you know much about him?

Yang Chen-Ning: Not much. My brothers know him well.

Ji Lizhen: Really?

Yang Chen-Ning: Because my brother and his family stayed in Kunming for a few years after I left. Did he go back to Tsinghua for a while and then go to Jilin University?

Ji Lizhen: Yes, he went to Jilin during the department adjustment and then to Dalian.

Yang Chen-Ning: He has made a lot of contributions to Jilin University. I have a student named Yu Lihua, who is very successful now. Yu Lihua is now 70 years old. He grew up at Tsinghua University and went to Jilin

University with his father after the department adjustment. His father, Yu Ruihuang, was transferred to Jilin University as vice president, so Yu Ruihuang knew Hsu Leetsch C. very well, and Yu Lihua also knew him. I think Yu Lihua's work could win a Nobel Prize. He works on the free electron laser. I asked Yu Lihua, and he said he remembered Hsu Leetsch C.

Ji Lizhen: Really? Okay. Thank you very much.

Yang Chen-Ning: It's nice talking to you, too. So, you can come and see me from time to time.

Ji Lizhen: Yes, I'll be back in Beijing sometime in March. Will you be here? I'll be back in Beijing in March of next year.

Wang Liping: Do you have time these days?

Yang Chen-Ning: Yes. Okay.

Ji Lizhen: Let's make another appointment.

The Sixth Interview

Time: August 8, 2018
Place: Institute for Advanced Study at Tsinghua University
Interviewer: Ji Lizhen, Wang Liping
Recorder: Wang Liping
Collator: Peng Cheng, Wang Liping, Ji Lizhen

As we all know, Southwest Associated University had many famous professors, such as Chern Shiing-Shen and Hua Loo-Keng, but there were also some teachers whose academic levels were not very high. Who were they? Southwest Associated University also trained many outstanding graduates in mathematics and physics. Who were they? What contributions have they made after graduation? What awards have they won? And what impact have they had? Naturally, there is romance at Southwest Associated University. For example, Yang Chen-Ning once experienced this kind of trouble when he was a student. How did he control his feelings? What tragedy did the rumors between Zhang Jingzhao and Hsu Pao-lu lead to? The story between Zhang Jingzhao and Hsu Pao-lu is, of course, quite different from that between Lin Huiyin, Liang Sicheng, and Jin Yuelin. In this interview, there are other stories, such as the story of Chang Iris and the book *The Rape of Nanking*, the discoverer of oracle bone inscriptions and the great mathematicians of his descendants. Why is the Yang–Mills theory more important than the symmetry theory of Yang Chen-Ning and Lee Tsung-Dao, which enabled them to win the Nobel Prize? Why did Dr. Yang say, "If I were a graduate student again today, I would do mathematics"?

Yang Chen-Ning: The biography Jiang Caijian wrote for me is a comprehensive and detailed description of my attitude toward being and doing things in many aspects and captures my attitude toward teaching students before 2000 when I was in my later years. It is a description from many angles. Just in these past two days, I was approached by a Chinese-American professor from the University of Michigan. I think he was from the School of Engineering. I don't know him, but he asked me for a book. I think there are about 50 Chinese professors now at your University of Michigan.

Ji Lizhen: Yeah, at least 300, maybe more. Because we have a Chinese Teachers' Association, there are hundreds of them.

Yang Chen-Ning: Is there one in Michigan State University?

Ji Lizhen: Michigan State University probably has more because we have a lot of people in the medical school who work in the labs there, and a lot of engineers.

Yang Chen-Ning: There's another person in your department whose early work I had something to do with. He's not Chinese. I can't remember his name. He's retired now. The two Chinese people I know best are in the physics department. A professor named Wu (Wu Qitai) retired and moved to San Diego, and the other one is named Yao Ruopeng. Do you know him? Yao Ruopeng also retired.

Ji Lizhen: That's right. We have the Ta-You Wu Lecture Series at the University of Michigan.

Yang Chen-Ning: That was established by Professor Wu. Do you still have Ta-You Wu Lecture Series? I was its first speaker back in the 1990s.

Difficulty in Translating a Biography

Ji Lizhen: What do you think about the matters regarding your translation and the translator? That is to say, how will the translation be handled? Do you have a suitable candidate for translating your biography? For example, someone you used to know.

Yang Chen-Ning: I think Jiang Caijian has this covered, right? I can put you in touch with Jiang Caijian.

Tsinghua University has a department of History of Science, which was newly established last year. This is another academic tradition in China. This department has a team of people working on the history of science and the philosophy of science. There was no one in this field at Tsinghua, but last year, or maybe the year before, they set up a department, and they invited an important person in this field, who was maybe about 60 years old. I went to a meeting when it was set up, so I've met this person. But my impression is that he's mainly engaged in science museums now, and he probably wants to have a science museum here at Tsinghua, so I don't think he's the most interested in the kind of history of science that we've just talked about.

Ji Lizhen: The thing is, it's okay for Chinese to translate from English into Chinese, while we're going to translate from Chinese into English now. There is some difficulty because people back in China are usually not good at English. They can read, but they can't write very well. For example, if we want to translate your biography into English for the foreign market, the English must be very authentic. So, my idea is something like this: we should find a Chinese whose English is also good, and first translate it into English, and then the second step is to find an American to polish the language.

Yang Chen-Ning: I think there is another way. For this reason, we need to train some Chinese people who can write English. In fact, there must already be many Chinese who can write English. The English department of Tsinghua University or Peking University should have a lot of good students, and these students are looking for a way out. What is their way out? They want to write novels. There are a few people who have become very famous there, which makes most of the young people who have just entered university think that this could be their future, too. The field we are now discussing is, in fact, unexplored. Just let me think about it.

Ji Lizhen: All right. For example, I bought this book the day before yesterday, *Thread of the Silkworm (the story of Qian Xuesen)*. You probably know this book; it was written by someone who grew up in America.

Yang Chen-Ning: I know this person very well. Her father earned a Ph.D. in physics from Harvard and was engaged in similar research areas as mine. After his Ph.D., he did some work at the Institute for Advanced Study, and then he was a professor at Ohio State University. Now he's retired. I think it was in the 1980s that he introduced his daughter Chang Iris to me. Chang Iris was going to write *Thread of the Silkworm (the story of Qian Xuesen)*.

Ji Lizhen: Did she interview you?

Yang Chen-Ning: She didn't just interview me. She asked me to give her all my materials about Qian Xuesen. I talked to her a little bit and gave her a book. It was in Chinese, I think, and it was already published in China at the time, but it wasn't officially published. I think it was written internally. It turned out that the book wasn't very useful to her. Then, she came to China by herself. Do you know her story after she arrived in Nanjing?

Chang Iris and *The Rape of Nanjing*

Ji Lizhen: Then, she wrote *The Rape of Nanking*.

Yang Chen-Ning: After she published *Thread of the Silkworm (the story of Qian Xuesen)*, she was not very successful. But when she came to China for this book, she became interested in the Nanjing Massacre and later wrote the book *The Rape of Nanking*, which caused a big sensation. And then, unfortunately, you know she killed herself?

Ji Lizhen: Yeah. Was she depressed or something?

Yang Chen-Ning: Her father's name is Chang Shau-Jin. He's retired now. I think in the last 30 or 40 years, he's moved on to other fields. *Thread of the Silkworm (the story of Qian Xuesen)* has never had much of an impact. *The Rape of Nanking* had a huge impact as soon as it came out, because the Japanese didn't like it very much, they didn't allow it to be published in Japan, so this is also …

Ji Lizhen: It just made her famous.

Yang Chen-Ning: It increased her fame.

Ji Lizhen: *The Rape of Nanking* is a great book, and her English is very good. I think she had a good beginning.

Yang Chen-Ning: She wrote it in English. I haven't even read it in Chinese. I just recalled this. You know there was a professor of mathematics at MIT whose name is Zheng Hong, whom I know very well. Do you know him?

Ji Lizhen: Is Zheng Hong a young man or an old man? What does he do?

Yang Chen-Ning: He's also 80 years old.

Ji Lizhen: What did he do?

Yang Chen-Ning: He wrote a book in English the year before last, and it's been translated into Chinese now. I think that it's called *Nanjing Never Cries* (published by MIT Press).

Ji Lizhen: Really? *Nanjing Never Cries?*

Yang Chen-Ning: He wrote it in English, and he probably translated it into Chinese himself. If you look it up, his English name is Zheng. This is his family name. His first name is Hong, so he is Hong Zheng. He's a professor of mathematics at MIT, and he was chairman of the department for many years, but he had an educational background in physics.

Ji Lizhen: He did math related to physics. Mathematical physics?

Yang Chen-Ning: Yeah, he's in the Physics Department of MIT, and the Mathematics Department of MIT has an applied mathematics section, which is influenced by Lin Chia-Chiao. He is there.

Wang Xiji and Yang Chen-Ning

Ji Lizhen: Yes, I was at MIT for three years. Lin Chia-Chiao, yeah, let me check. There is another person who is also from Southwest Associated University. I saw another book called *Great Masters* the day before yesterday. There is a person who graduated from Southwest Associated University and this person was singled out in the book.

Yang Chen-Ning: That's the guy.

Ji Lizhen: Do you know this person? He was from Yunnan Province. He went to Southwest Associated University. He wrote about Southwest Associated University very well. What do you think of his writing?

Yang Chen-Ning: Wang Xiji. He's still alive and is about the same age as me.

Ji Lizhen: Really? Is he still alive?

Yang Chen-Ning: He is one of the 23 national heroes of China (winner of "Two Bombs and One Satellite Meritorious Medal"). I know of him, but I'm not that familiar with him.

Ji Lizhen: One winner of "Two Bombs and One Satellite Meritorious Medal."

Yang Chen-Ning: He was in my class at Southwest Associated

University. I didn't know him at that time because he was in engineering college, and I didn't meet him until recently. Now, there are only three people alive from my class at Southwest Associated University: he and I, and another named Xu Yuanchong. Xu Yuanchong is a professor at Peking University, and he is famous as a translator. There is a celebration in Kunming on November 1st of this year, so I guess they will probably invite both. Wang Xiji is an ethnic minority from the Bai ethnic group in Yunnan. He is a typical Chinese in behavior and attitudes, not arrogant and quite popular among the people around him.

Xu Yuanchong and Wang Xiji

Yang Chen-Ning: Xu Yuanchong has a loud voice. Wang Xiji is not well known. Although he is one of the 23 people in China—do you know the names of these 23 people? After a long discussion in the State Council of China, the 23 people were finally announced around 2000. Among them, there were those who made atomic bombs, missiles and satellites. He was one of those who made satellites.

Ji Lizhen: He said he was surprised when he went to the meeting that day and they gave him an award.

Yang Chen-Ning: When was this?

Ji Lizhen: His article said that when he went to the Great Hall of the People for a meeting, he was awarded a prize.

Yang Chen-Ning: Is that the time when someone interviewed him?

Ji Lizhen: Yes, he was interviewed by CCTV.

Yang Chen-Ning: Wait a minute.

Ji Lizhen: This was part of a CCTV program.

Yang Chen-Ning: They also visited Ting Samuel C. C.

Wang Xiji and Southwest Associated University

Ji Lizhen: Southwest Associated University was mentioned in the TV program.

Yang Chen-Ning: As for the book of *Great Masters*, if there is volume one, there must be volume two. Xu Chen, can you please order a copy of this book, *Great Masters*, the first and second volumes?

Ji Lizhen: Yes, there are two volumes. I checked and there seem to be only two volumes. Volume One and Volume Two. When he was asked about his college experience, he specifically wrote about Southwest Associated University. Please tell me how you felt at that time because this is what he wrote about Southwest Associated University.

Xu Chen: I can't find this book.

Ji Lizhen: Maybe it was published earlier. I bought it at an old bookstore inside Tsinghua University.

Yang Chen-Ning: It was by the Commercial Press.

Ji Lizhen: What do you think of what he said? He was obviously very affectionate and impressed with Southwest Associated University because he said that there was no other university like Southwest Associated University in the history of the world.

Academic Level of Teachers at Southwest Associated University

Yang Chen-Ning: Here's the thing. It is difficult to explain clearly how high the academic standards of the teachers at Southwest Associated University were at that time. If you want to say so because of him, Southwest Associated University at that time was very successful. That is absolutely true. If you ask what the reason for the success was, then it becomes a very complicated question. If he gives you the impression, or Wang Xiji gives you the impression, that the academic standards of the teachers at Southwest Associated University were the best in the world …

Ji Lizhen: Yes, there were many great masters.

Yang Chen-Ning: Well, maybe not exactly.

Ji Lizhen: Really?

Yang Chen-Ning: In the field of the literature at that time, I think it should be said that there were some people who seemed to occupy an important place in the history of Chinese literature later, such as Shen Congwen and Zhu Ziqing. I think Zhu Ziqing should be counted as one of them. Therefore, in the history of Chinese literature, the teachers at Southwest Associated University did occupy a very important position. The reason for me to somehow disagree is maybe Shen can occupy a seat, but there was also a drama writer named Sun, and I don't think they both reached the first-class level. When it comes to science, I think Southwest Associated University had nothing to doubt about. It had three first-class teachers in mathematics, namely Hua Loo-Keng, Chern Shiing-Shen and Hsu Pao-lu. What does it mean to be first-class? If you take today's standards, of course, you have to become a member of the American Academy of Sciences or the Royal Society of Britain. If you take this as first-class, then do they meet the criteria for the Fields Medal? If you ask me, I think Chern has met the standard, but I dare not tell my opinion of Hua and Hsu. If it is said that they could easily be elected as members of the American Academy of Sciences, then I think all three of them had reached this level, and no one else in science had at that time.

Ji Lizhen: What about Chung Kailai, who came later? Did Chung Kailai also study at Southwest Associated University, or did he teach?

Yang Chen-Ning: That's a good question, but Chung Kailai was a student at that time.

Ji Lizhen: A student, so not at such a high level yet?

Yang Chen-Ning: I think so. If you ask me, he should have become an academician at the American Academy of Sciences. But he never did. In my mind, he never did because he had problems with his character.

Chung Kailai's Temper

Ji Lizhen: He had a bad temper and liked to swear a lot.

Yang Chen-Ning: Nobody liked him.

Ji Lizhen: Yeah, he liked to swear a lot. The first time I met him, he seemed to swear quite a bit.

Yang Chen-Ning: You met him?

Ji Lizhen: Yeah, I met him several times

Yang Chen-Ning: Oh, do you have anything to do with his field?

Ji Lizhen: What I used to do was a little bit related to harmonic functions.

Yang Chen-Ning: He went to America on the same ship with me.

Ji Lizhen: You were on the same ship?

Yang Chen-Ning: There were 20 people on the same ship, and we didn't like him.

Ji Lizhen: Because he seemed to be irritated and judgmental.

Yang Chen-Ning: I think he's good at math, but he's not popular. There's a guy named Ky Fan. Do you know him? I have the impression that he did very well, but few Chinese know about him.

The Ky Fan Fund of the American Mathematical Society

Ji Lizhen: Yes, I do know of him because now the American Mathematical Society has set up a Ky Fan Fund. His English name is Ky Fan. He donated money in his family's name to set up a foundation to help Chinese and American scholars do exchanges. I'm on the committee of the foundation.

Yang Chen-Ning: He was in the United States later.

Ji Lizhen: Yes, in Santa Barbara or somewhere.

Outstanding Southwest Associated University Mathematics Department Students

Yang Chen-Ning: Later, (Ky Fan) was also very unhappy. Finally, I think he went to Chinese Taiwan for a few years. Chern Shiing-Shen admired him very much. In the Department of Mathematics at Southwest Associated University, there were several successful mathematicians. One was Liao Shantao, and the other was Wang Hsien Chung, who did group representation theory.

Ji Lizhen: Wang Hsien Chung later went on to Cornell University.

Yang Chen-Ning: Wang Hsien Chung is a professor at Ithaca.[58]

Ji Lizhen: Yeah, he went to Cornell University.

Yang Chen-Ning: Chern Shiing-Shen was very appreciative of him, too.

Ji Lizhen: Yes, I know some of his achievements because he is related to

[58]Cornell University.

discrete subgroups of Lie groups, homogeneous spaces and symmetric spaces. There is now an assistant professor title at Cornell University named after him.

Yang Chen-Ning: What's his name?

Ji Lizhen: It is called the H.C. Wang Assistant Professor.

Yang Chen-Ning: You mean it was established to honor him?

Ji Lizhen: Yeah, and then the postdoc title was also named after him.

Yang Chen-Ning: He and I were probably in the same class at Southwest Associated University.

Wang Yirong, Oracle Bone Scripts and Wang Hsien Chung

Ji Lizhen: Really? You two were in the same class?

Yang Chen-Ning: He is the grandson or great-grandson of Wang Yirong. Wang Yirong was the libation performer of the Imperial College at the time of the Eight-Nation Alliance, who served an equivalent role to the chancellor of Peking University today. His whole family committed suicide by jumping into a well when the Eight-Nation Alliance (invaded the capital). It is possible that Wang Hsien Chung's father was not in Beijing at that time, so he did not die. At least Wang Hsien Chung had not been born at that time. Then Wang Yirong made a very important contribution; he discovered oracle bone inscriptions.

Ji Lizhen: Really? He found them when he bought some Chinese medicine.

Yang Chen-Ning: It was widely reported that Wang Yirong was an important Sinologist at that time, and he was the libation performer of Imperial College. One day, he fell ill, and the doctor gave him a prescription. He sent his servant to buy medicine according to the prescription, and when he came back, he said, "Let me see." One of the prescriptions was a tortoiseshell—just pieces of tortoise shell. He saw that some of the pieces had engravings on them. He was an expert on inscriptions on ancient bronze objects, so when he looked at these, he thought it looked like the inscriptions on ancient bronze objects, but he didn't know for sure, so he thought it might have something to do with ancient Chinese characters. He sent his servant back to that shop to do three things: first, to buy all the tortoiseshell; second, to ask where it came from; third, to tell the servant not telling anyone else. The servant then went away, and the first task he did correctly; as regards for the second, he was told it came from somewhere in Shandong, but as for the third instruction, he disobeyed and told others everywhere. So pretty soon, everyone in Beijing heard about it and thought they would also find ancient Chinese characters, so everyone was buying them up like crazy. At that time, as long as there was a symbol on the turtle shell, it could be sold for one *liang* (a unit of currency weight equivalent to 50 grams) silver, which was very expensive. Then, they sent people to Shandong, where there was nothing at all, just to cheat them out of their money. But anyway, it had started, and finally he found out that it was in Anyang. So, he sent someone to Anyang, where Wang Yirong's family bought a lot of Anyang's things. Later, after he died, this group of Wang's items was sold to Liu Tieyun. It was Liu Tieyun who wrote *Travels of Lao Can*. So, Liu Tieyun turned the book he bought from the Wang family into rubbings and published it. It was the first published book about tortoiseshell in Chinese. It was called *Tie Yun Cang Gui*. It's historically famous because it was later proven to be the oldest writing in China. I once asked Wang Xianzhong if he kept a copy at home, and he said not anymore.

Ji Lizhen: I don't know about this. I didn't know that his ancestors found oracle bones. Mr. Wang's book is also very good. There is another Chen Kuo-Tsai. Do you know him? Chern Shiing-Shen also really admires him.

Yang Chen-Ning: What's he called in Chinese?

Ji Lizhen: 陈国才 (Chen Kuo-Tsai). He studied in topology at the University of Illinois at Urbana-Champaign. Sometime around 1990, Chern Shiing-Shen held a meeting for him in Nankai (memorial meeting, 2000). He seemed to have done original and good mathematics. Did he go to Southwest Associated University, too?

Yang Chen-Ning: That's right. There was another very successful student named Wang Hao.

Ji Lizhen: Oh, Wang Hao, who specialized in logic.

Yang Chen-Ning: So Chung Kailai, Wang Hao and Liao Shantao.

Ji Lizhen: Yes, were you all classmates?

Yang Chen-Ning: Liao Shantao was in my class. My impression is that after Liao Shantao came back to China, he was very successful. So if you ask about the teachers of the mathematics department of Southwest Associated University, with the exception of Chern, Hua and Hsu, I don't think they did much really … Perhaps Jiang Zehan did, but his main work was not in the topology of contributions; however, he trained some (notable) Chinese students. Duan Xuefu and Cheng Minde were a little older than me, and my impression is that they did very well later on. If they were in the United States, they did not reach the level of academicians. Both Duan Xuefu and Cheng Minde later became important figures at Peking University.

Ji Lizhen: Yes, Hsu Pao-lu didn't seem to have many students under his name. Hsu Pao-lu seemed to have a good knowledge foundation. Chern and Hua seemed to have trained a lot of people who came afterwards.

Hsu Pao-lu and Other Students of Southwest Associated University

Yang Chen-Ning: Hsu Pao-lu came to Southern California after the victory of the War of Resistance against Japanese Aggression and set up a center that still exists today which has been very important in the field of mathematical statistics internationally. He was the founder. He didn't stay there long enough to adapt to foreign life, so he went back to China.

Ji Lizhen: Oh, that's the reason. Did it make a big difference when he came back?

Yang Chen-Ning: After he went back to China, he was highly valued, but he never got married. His sister was Yu Pingbo's wife, and I knew Yu Pingbo's son very well. Yu Pingbo had two daughters and one son, both of whom I knew very well. The two daughters were in my primary school class. I have known Hsu Pao-lu since I was a child. I even attended his classes at Southwest Associated University. After returning to China, he was not very sociable. I had the impression that he did not seem to connect with anyone. The mathematics circle and its students probably respected him very much. Later, he committed suicide during the "Cultural Revolution."

Ji Lizhen: Really? A "Cultural Revolution" suicide.

Yang Chen-Ning: He committed suicide during the "Cultural Revolution." Do you know the story? Zhang Jingzhao was a female student who majored in mathematics in my class at Southwest Associated University. Xu Chen, please, let me see, can you give me a copy of my *Forty Years of Reading and Teaching*, and also hand me the *Dawn Collection*. Xu Chen, excuse me, but I have to ask you to find me another book, *Selected Papers*. I can't find the photo right now.

Yang Chen-Ning and His Female Classmate Zhang Jingzhao

Ji Lizhen: A female mathematics major?

Yang Chen-Ning: When I was a sophomore at Southwest Associated University, I had a female classmate majoring in mathematics named Zhang Jingzhao. I had a crush on her for a while. She was about a year or two older than me, and I have a picture of her. I'll tell you when I find it. I'll email you. Then, after a few months like this, a semester, I myself …

Ji Lizhen: You were not annoyed.

Yang Chen-Ning: Yes, I was.

Ji Lizhen: Why?

Yang Chen-Ning: I thought it was very bad for me.

Ji Lizhen: Why was it bad?

Yang Chen-Ning: I still remember writing it down later. I did a self-reflection and thought that my emotional world was like a very calm lake, but after Zhang Jingzhao came, it turned out to be very choppy, so I made a decision.

Ji Lizhen: To concentrate on studies?

Yang Chen-Ning: To treat coldly the relationship, so I became indifferent, and from then on, I decided to keep (a distance) from her. Because of my father and my mother … well, she was a mathematics major, and since my father was a mathematics professor, she sometimes came to my house to choose courses, so my father and my mother liked her very much. Therefore, she maintained a very good relationship with our family, but later I did not date with her. After she graduated, she married an economics graduate named Wang Chuanlun, and then she went to Guizhou to teach in a middle school. After the victory of the War of Resistance against Japanese Aggression, she became a teacher at Peking University. She has been working in the mathematics department of Peking University since then.

Ji Lizhen: Really? What math did she do later?

Yang Chen-Ning: I don't think she did any important research, but she was in the mathematics department at Peking University. She probably made a lot of contributions in her early years, and she always had a very close and good relationship with my family. For example, in 1957, my little brother, Yang Zhenfu, was a freshman in the Department of Mathematics at Peking University. He had schizophrenia, so my father wrote to Zhang Jingzhao and asked her to send my brother back home to Shanghai. So, it was Zhang Jingzhao …

Ji Lizhen: She took your brother home.

Zhang Jingzhao and Hsu Pao-lu

Yang Chen-Ning: Zhang Jingzhao was a teacher at that time. She accompanied Yang Zhenfu by train all the way to Shanghai, and thus she maintained a good relationship with my family. After Hsu Pao-lu came back to China, he became a professor at Peking University. He had no wife, and he could not take good care of himself. Zhang Jingzhao had been in his class at Southwest Associated University, so Zhang Jingzhao asked Hsu Pao-lu to live in their home. Her husband's name was Wang Chuanlun. I don't think Wang Chuanlun was a professor at Peking University yet at that time. He was a professor at some economic and trade university. As a result, during the "Cultural Revolution," there were rumors that Zhang Jingzhao had an illicit relationship with Hsu Pao-lu. Later, both of them committed suicide. When I came back in 1971, both of them were gone, and I heard that both of them had committed suicide, but I didn't inquire further. I didn't go … In 2000, after I returned to Tsinghua University, there was a reunion meeting at Southwest Associated University. Wang Xiji, Xu Yuanchong and Wang Chuanlun were all there. I saw Wang Chuanlun, Zhang Jingzhao's husband, so I met him once. At that reunion, we had a meal, and I didn't ask what happened to Zhang. Then Wang was gone, and of course, he is dead now. For the history of Hsu Pao-lu, and what exactly happened to Hsu Pao-lu in this period, I do not believe that they had any type of improper relationship, as had been rumored at that time. I think this cannot be correct according to my understanding of them two. I think it was impossible. But I haven't seen anyone write about this thing. I wonder if Hsu Leetsch C. wrote about it. You can write to him and ask.

Ji Lizhen: Yes, I'll ask him.

Yang Chen-Ning: Let me see. Who else might know?

Ji Lizhen: I don't think anyone else did. I don't think there's anybody there anymore.

Yang Chen-Ning: Just let me think about it. Wang Xiji may have some idea. After all, Wang Xiji was at Southwest Associated University, and he was later in Beijing.

Ji Lizhen: Does Zhang Jingzhao have any children?

Yang Chen-Ning: Zhang Jingzhao? I don't know.

Ji Lizhen: You don't know? Because I think if she had a baby, she might have felt more …

Yang Chen-Ning: Yes, her husband's name is Wang Chuanlun. He graduated from the Department of Economics at Southwest Associated University.

Lin Huiyin, Liang Sicheng and Jin Yuelin

Ji Lizhen: I heard that there was another similar situation at Tsinghua University. There was a famous architect named Liang Sicheng, and his wife was Lin Huiyin. Wasn't she the one Xu Zhimo fell in love with?

Yang Chen-Ning: Yes, that's Xu Zhimo. They were older than me, so it was a little bit earlier.

Ji Lizhen: Because later people told me, and I also read in a book, that there was a professor of philosophy at Peking University who also loved Lin Huiyin very much. Later, the three of them all lived in the same courtyard.

Yang Chen-Ning: Jin Yuelin.

Ji Lizhen: Was this story true? Did the three of them live together?

Yang Chen-Ning: There were a lot of articles written about it. I don't know. It was said that they were very close with each other.

Wives of Professors at Tsinghua University

Yang Chen-Ning: Liang Sicheng's wife was a very outgoing woman. Speaking of this, have I ever told you that when I was a child, I lived at Tsinghua University for eight years? We lived in Xiyuan then. If you asked me about that time, I think there were more than 50 professors, but less than 100.

Ji Lizhen: Really? That's not many in size.

Yang Chen-Ning: The professors' social relationships were divided into three categories, depending on whether their wives were more social. One group, my mother's group, didn't have much education. Next group went to, say, university in China. And then there was the last group who had studied abroad. Apparently, the three groups of families had little contact with each other, and members of each group had more contact within itself. For example, my mother was very familiar with Xiong Qinglai's wife because they were about the same age; they used to have bound feet, later called liberated feet, and they had more contacts. Another group, like Zhou Peiyuan's wife, was made up of wives who attended university. Liang Sicheng's wife was one of the few who had studied abroad, so they formed a completely different social group. I haven't seen anyone write an article about it. In fact, it's also very meaningful, especially for the children of these people who grew up in Tsinghua University at that time, and there were a lot of people like me, a little bit younger than me. A few years ago, they published a series of books that were like a compilation of several picture books about the different families in Tsinghua University before the War of Resistance against Japanese Aggression. The main person who did this was Jiang Zehan's son, whose name seemed to be Jiang Pidong. If you check this out, the library of Tsinghua University must have these books somewhere. If you look under Jiang Pidong, 江丕栋 in Chinese, you will find books like this.

Ji Lizhen: Yes, I think Tsinghua University used to have a better feeling and atmosphere. For example, the environment was a little better than now. I wonder if the teachers there used to be better, because now there are more universities, and they are scattered. My previous personal feeling was that Tsinghua was more special.

Yang Chen-Ning: Part of the reason is that Tsinghua is so big now. Did Jiang edit this?

Ji Lizhen: No. Zhou Wenye did.

Yang Chen-Ning: I made a mistake. It was edited by Zhou Wenye.

Ji Lizhen: Zhou Wenye, yes. The other reason that Southwest Associated University was so successful was that the quality of students was very high.

Good Times for Southwest Associated University Graduates

Yang Chen-Ning: One thing that I think people didn't talk about Southwest Associated University is that its timing was right. After the founding of the People's Republic of China, the graduates of Southwest Associated University became important contributors to the 17 years before the "Cultural Revolution." These people had this opportunity in different fields because it was tantamount to a country starting from scratch. A country starting from scratch requires a lot of knowledgeable people, many of whom came from Southwest Associated University. Of course, you may ask, since there were other schools at that time, whether other schools had students as good as Tsinghua University or Southwest Associated University? Because at that time, the best students in the country were at Southwest Associated University, and that's for sure. The only other school that was slightly inferior was the Central University in Chongqing. I think the graduates of these two schools contributed greatly to the success of these two schools for the 17 years after liberation and the founding of the People's Republic of China.

Ji Lizhen: Yes, it's true that times make heroes. I have another question. You just said that when you were at Southwest Associated University, you fell in love with Zhang Jingzhao for a period of time, and then you thought that after thinking it over …

Yang Chen-Ning Breaks Off His Feelings for Zhang Jingzhao

Yang Chen-Ning: I don't think it could be called falling in love, but I think it had a great influence on me, and on Zhang Jingzhao …

Ji Lizhen: Did she know you liked her?

Yang Chen-Ning: Obviously, of course. But I don't think it affected her as much as it affected me during those few months. I was relatively immature, and she was maybe a year or two older than me, and I think, given her attitude and everything, that she was a very mature person.

Ji Lizhen: How old were you, as a college student? About 20?

Yang Chen-Ning: We were sophomores.

Ji Lizhen: So 19 in the second grade?

Yang Chen-Ning: I was 17 years old. I went to college when I was 16.

Ji Lizhen: Seventeen. I think you were quite amazing. How could you withhold emotion to a girl you like? However, after you thought it over and made up your mind, you thought you should be calm, and I don't think many people can do that.

Yang Chen-Ning: I wrote it all down. When I get back, I will find it out and send you an email.

Ji Lizhen: I think it's not easy. Most people can't do it. Because if a guy really likes a girl, if he's really into her, it's hard to calm down.

Yang Chen-Ning: I think this is coming up again … Because if you have read the *Weiyang Song*, or if you have read Xu Yuanchong's autobiography, you will know that many of my classmates at Southwest Associated University at that time took a different path from me. You could say I was a good student brought up by traditional Chinese education.

Ji Lizhen: Be able to control your feelings.

Yang Chen-Ning: One with strict self-discipline.

Ji Lizhen: Yeah, I don't think many people can do that, and I don't think so now, either, because I think there are a lot of men and women out there who, because their relationships have had a big impact on their lives, find that it's not easy to do that. You said just now that there were a lot of people who made big achievements in the mathematics department, but what about the physics department? Are you two the only ones in the physics department? You and Lee Tsung-Dao?

Southwest Associated University Physics Department Students

Yang Chen-Ning: In the physics department, for example, some of my classmates were very important. One is Dai Chuanzeng, who later served as the director of an institute of the Academy of Sciences, near Beijing, in Zhoukoudian. The institute is still there, and it is a large place for the study of nuclear physics, which he later worked on. Another one was Xiang Rensheng, who later went to the United States to study. After he returned to China, I remembered that he had made important contributions to the development of acoustics (microwave magnetism) in China. There was also a student named Huang Zuqia, who graduated from the physics department of Southwest Associated University. He was very important and was junior to me. He was in the same class as Lee Tsung-Dao from 1945 to 1946. Later, he was an important figure in China's atomic bomb project, but he was not among the 23. There was a famous event in China's atomic bomb industry, and he was very instrumental in it. You will have to ask the Chinese atomic bomb people what the actual story was. All in all, he made a major contribution to China's defense industry, second only to the 23 people. At that time, there was also a man named Lu Zuyin, who was in the same class as Lee Tsung-Dao, Huang Zuqia, and Ye Minghan. However, their status in the international academic field of physics was not that much. Huang Zuqia, for example, has made great contributions to China, but in the field of application, in the international physics academia, in fact, he was not a significant figure. Another important figure who was not from Southwest Associated University but from Tsinghua University in the early years was Peng Huanwu, who became an important figure. Peng Huanwu was one of the 23 heroes. He had returned to China. I think he was about five or six years older than me. When I was at Southwest Associated University, he came back to China and was a professor at Yunnan University. Later, he was an important figure in the development of the atomic bomb in China. In the 1930s and 1940s, everyone paid much attention to these people. For example, there were famous men who studied physics at Tsinghua University first and at Peking University later, such as Wang Zhuxi and Zhang Zongsui. Do you know Zhang Zongsui?

Ji Lizhen: I don't know him.

Yang Chen-Ning: You haven't heard of Zhang Zongsui?

Ji Lizhen: Was he in mathematics or physics?

Yang Chen-Ning: Zhang Zongsui graduated from Tsinghua University. He was ten years older than me. And then there was Wu Ta-You.

Ji Lizhen: Oh, I heard about that.

Yang Chen-Ning: And Ma Shijun, Peng Huanwu, Lin Chia-Chiao …

Ji Lizhen: I've heard about Lin Chia-Chiao, too.

Yang Chen-Ning: These six or seven people were the most famous among the new generation in physics and mathematics in the 1930s.

Ji Lizhen: They were from Tsinghua University. What about Southwest Associated University?

Yang Chen-Ning: Most of them were from Tsinghua University, and some came from Peking University.

Ji Lizhen: Were some from Southwest Associated University?

Yang Chen-Ning: Yes, one of them was Peng Huanwu. Peng Huanwu graduated from Tsinghua University in Beijing in the 1930s. He later went to England. During the War of Resistance against Japanese

Aggression, he spent a lot of time at the University of Edinburgh in Scotland, UK. He was very familiar with Schrödinger at that time in Dublin, Ireland, and also with William Rowan Hamilton. Peng Huanwu was very famous when I went to study in the United States because he put forward a theory called HHP (Hamilton–Heitler–Peng theory) with Hamilton, and he was very famous when I was studying in Chicago, but their theory didn't succeed. So, Peng Huanwu did not make any important contribution in the academic field of physics for a long time, but he made an important contribution to China's atomic bomb work, so he was one of the 23 heroes. If you want to ask who were trained in Southwest United University in the 1930s, then I think he would certainly be considered one of the important figures.

Some articles were written about Zhang Zongsui, but it is actually worth someone writing a biography. I think Zhang Zongsui was very clever. I don't know him, but I think he is a bit arrogant. I think he was born around 1912–1914. Why do I know that? After he graduated from Tsinghua University, he went to England to do graduate work and soon got a Ph.D. He became good friends with Dirac when he was in England, so he always had a very good relationship with Dirac. Then he went back to China. After the victory of the War of Resistance against Japanese Aggression, he went back to China and became a professor at Peking University. There are still some of his students around who retired. His father, Zhang Dongsun, was a famous philosopher, so he and his father ... How do we know that he was very clever in England? Because he was a good friend of Dirac, he was ten years younger than Dirac, and he soon got his doctorate. He had the same doctoral supervisor as Mr. Wang Zhuxi, so his articles and Mr. Wang's articles were all about statistical mechanics. Statistical mechanics was not the most popular science at that time. The most popular science was field theory, and he also practiced field theory. Therefore, after he finished his Ph.D., he was recommended by Dirac to go to the institute of Bohr in Copenhagen. In the 1930s, Bohr's institute was the center of cutting-edge physics in the world, and he became very familiar with Bohr and his son there, and then he went back to Copenhagen before the liberation, after the victory in the war. In this way, he was different from Mr. Wang. Since his return to China, Mr. Wang had nothing to do with the most popular field at that time: field theory. He has always stayed in statistical mechanics. But Zhang Zongsui gave up statistical mechanics and devoted himself to field theory. That's why I read his articles on field theory in the 1930s. Therefore, when I went to the United States, if you want to ask which young people in China were the most famous in physics and in the frontier of theory, the two most famous people at that time were Zhang Zongsui and Wang Shoujing, who was about ten years older than him. Do you know this man, Wang Shoujing?

Ji Lizhen: I don't know him. Did he also graduate from Tsinghua University?

Yang Chen-Ning: Wang Shoujing, written as 王守竞 in Chinese. Wang Shoujing, I think, wrote an essay when he was studying at Columbia University at the very beginning of quantum mechanics in 1926–1927, and he became one of the most famous physicists who studied for a doctorate at a very young age.

Ji Lizhen: That was very promising.

Yang Chen-Ning: Very promising. Because when quantum mechanics first started, Pauli and Schrödinger solved a problem, which you must have read, about the harmonic oscillator. Schrödinger's equation could solve just about anything like this with this harmonic oscillator, except for one problem, which was not solved and was finally solved by Wang Shoujing. Thus, he immediately became the most famous young man in cutting-edge physics at the time. A man who was studying at Columbia University at the same time was called Rabi, and he became an American Nobel Prize winner. Rabi was an experimental physicist and the leading voice of American physics in the 1950s. He was in the same class as Wang Shoujing and admired him very much. After Wang Shoujing returned to China, I did not know where he was. After teaching for a year or two, he was invited by Chiang Kai-shek to work in a munitions factory. So, later, during the War of Resistance against Japanese Aggression, he was the director of the munitions factory.

Ji Lizhen: Then he stopped doing physics.

Yang Chen-Ning: Since then, he stayed in government with a very high position and was doing this technical work. He didn't return to China after liberation. He stayed in the United States. There was also a strange thing about him staying in America that I never understood. At that time, America was developing, so universities were asking for professors. If he wanted to become a professor at a university, I think Rabi would immediately have recommended him, but he didn't go. So, instead of staying in physics, he ended up in a research institute at MIT, where he worked as a programmer, and he spent the rest of his life as a programmer. He also had several younger brothers who made great contributions in China. Two younger brothers, Wang Shoujue and Wang Shouwu, were, successively, the director of the Semiconductor Institute. The Wang family was very prominent, and the brothers and sisters were all very well known in the educational circles of China. In addition to Wang Shoujing, Wang Shoujue and Wang Shouwu, they had another sister. A few years ago, she passed away at Tsinghua University. She was over 100 years old, and her name was Wang Mingzhen. For many years, the most important, most famous and internationally renowned professor of physics at the University of Michigan was Uhlenbeck. There were two very good Chinese female doctoral students: Wang Mingzhen and Wang Chengshu. After these two returned to China, they both had quite good status in the physics academia, and they were Uhlenbeck's best postgraduate students. Wang Mingzhen, on the other hand, became a professor at Tsinghua University for a long time. She passed away at about 100 years old. The second one, which I know very well, made great contributions to China. I think she should have been one of the 23 contributors, but she was not. She is one of the people who made great contributions to atomic isotope separation in China.

National Science Awards Received by Southwest Associated University Students

Ji Lizhen: That's not easy. I have another question. You said that

Southwest Associated University was good at mathematics. Also, several people won the science prize: Hua Loo-Keng won the first award, Hsu Pao-lu won the second, and Chung Kailai won the third. What was the social impact of the awards at that time?

Yang Chen-Ning: They had a very big effect.

Ji Lizhen: What about the media reaction? The media is quite developed now.

Yang Chen-Ning: Hua didn't have a higher education, so all of a sudden, he became very successful and this made people pay special attention to him.

Ji Lizhen: Because he won that award, he became very famous, didn't he?

Yang Chen-Ning: He succeeded without higher education, which was admired very much in China. The United States admired it, and China admired it even more.

Ji Lizhen: That's right. He's totally self-taught. That's not easy.

Hua Loo-Keng Is a Genius

Yang Chen-Ning: He's a genius. There's no doubt about it.

Ji Lizhen: At that time, winning a science award had a great impact on society. All the media covered the event.

Yang Chen-Ning: Yes, and he was not only influential domestically but also internationally. For example, in the Soviet Union, Ivan Matveyevich Vinogradov paid a lot of attention to him, so he was specially invited to the Soviet Union.

Ji Lizhen: I read a biography the other day, and it said that they had decided to give Hua Loo-Keng a Stalin Prize, or whatever. But he didn't get it, because Stalin died.

Yang Chen-Ning: If you ask about the Soviet Union wanting to give an award to a Chinese in Stalin's time, I think it would have to have been to Hua Loo-Keng.

Ji Lizhen: But no one won a prize in physics?

Yang Chen-Ning: No, and no one had made one for physics. In terms of physics, I was talking about Zhang Zongsui, Wang Shoujing, Wang Zhuxi and Peng Huanwu. They were not so high up in the international physics circle.

Ji Lizhen: How about Wu Ta-You? It seems that Wu Ta-You has had a great influence.

Yang Chen-Ning: Wu Ta-You won an award.

Ji Lizhen: He won a science award?

Yang Chen-Ning: I don't know what kind of award he won. Yes, if you asked the Chinese physics circle in 1945 who held a position in the international physics circle, I think Wu Ta-You would be the one they would mention. The others would be Peng Huanwu, Zhang Zongsui and Qian Xuesen.

Ji Lichen: Qian Xuesen.

Yang Chen-Ning: I think they were all quite famous.

Ji Lizhen: Yeah. There's another problem. The Fields Medal was announced last week. You've heard of it, the World Congress of Mathematicians was held every four years, so it was announced last Monday, and when it was announced, a lot of people were a little surprised and even a little disappointed because the Fields Medal seemed to be dropping in quality.

Yang Chen-Ning: How many recipients were there?

Deteriorating Quality of the Fields Medal and the Nobel Prizes

Ji Lizhen: Yeah, several recipients were not expected. A lot of people I've met were disappointed. I think the quality of the Nobel Prize is better. Do other people feel surprised or disappointed when Nobel Prize winners win the prize? Or has the quality of the Nobel Prize always been better?

Yang Chen-Ning: Please say more. What's people's reaction to the four Fields Medal winners in mathematics?

Ji Lizhen: There was one person who everyone thought was absolutely fine, admired and recognized.

Yang Chen-Ning: Didn't he get it?

Ji Lizhen: Yes, he got it.

Yang Chen-Ning: Yes.

Ji Lizhen: So, there's a problem with the other three.

Yang Chen-Ning: The other three are a bit confusing, aren't they?

Ji Lizhen: Yeah, why were they chosen? There was a guy who solved some problems in a whole new way, and people thought he should take it. The others did not do a good job.

Yang Chen-Ning: How many Americans were there in these four?

Ji Lizhen: None. One German, one Italian, one Iranian, and one …

Yang Chen-Ning: There's an Iranian one?

Ji Lizhen: Yes, Iranian.

Yang Chen-Ning: Where is the Iranian?

Ji Lizhen: He is in England now. Oh, and an Indian, now living in the United States.

Yang Chen-Ning: Do any of them work on string theory?

Ji Lizhen: No.

Yang Chen-Ning: So, you are saying that there is one person who everyone agrees should win, and three others who are disputed.

Ji Lizhen: People are surprised. A lot of people feel as if the quality of the prize is going down. It seems that the Nobel Prize's quality is not bad. The Nobel Prize selection was quite carefully, or the judging panel was very careful.

Yang Chen-Ning: I think the Nobel Prize in Physics has also had this phenomenon.

Ji Lizhen: Also?

Nobel Prize Problems

Yang Chen-Ning: But the nature of that is a little different. I'll give you an example. About five or six years ago, they suddenly gave out a prize because they thought that a new discovery could make a very significant contribution. It's amazing to have a new development in the field of new materials, so a lot of people thought it had a great future, but most of them were still in a wait-and-see mode. After giving it to these two or three people, most of the physicists were a little surprised.

Ji Lizhen: What is it, the Nobel Prize in Physics?

Yang Chen-Ning: Yes, the Nobel Prize in Physics, but it didn't get much discussion. But now, after a few years, it has been recognized that the original enthusiasm was a little early, so now most people think that the prize …

Ji Lizhen: Was it wrongfully awarded?

Yang Chen-Ning: Either that it should not be given, or at least not too early. I think this is unique to physics … So, to get to that, let me tell you about one that relates to me. You know, if you go to this Nobel Prize website, https://www.nobelprize.org/, it has an application, and when you go to that application, you can ask who recommended Hoft for the Nobel Prize, and it will list it for you. You can also ask when and whom Hoft recommended, too. So the other day I asked Xu Chen to search there asking who recommended Yang Chen-Ning.

Ji Lizhen: Yeah, I think that's a good question.

Yang Chen-Ning: It was blocked.

Ji Lizhen: Is it? Why is that?

Yang Chen-Ning: Who recommended Yang Chen-Ning? The information was blocked.

Ji Lizhen: What's the reason? Is it old?

Yang Chen-Ning: Later, I asked a person inside the Nobel Foundation, and he said that they had blocked this information about Yang Chen-Ning until all the people who were related to Yang Chen-Ning were gone. So you can look it up in the future. You can go look it up after I'm gone.

Ji Lizhen: So, he's actually …

Why Didn't Yang Chen-Ning Win a Second Nobel Prize?

Yang Chen-Ning: I wonder why, because I think it's a strange thing in physics that I didn't win a second Nobel Prize.

Ji Lizhen: Yes, Yang–Mills is very important.

Yang Chen-Ning: So, I don't think they blocked Lee at this point. At least Lee wasn't completely blocked.

Ji Lizhen: So who makes the decision, to block or not to block?

Yang Chen-Ning: The Nobel Foundation itself.

Ji Lizhen: Oh, the Nobel Foundation itself. Yes, I think the influence of Yang–Mills should be much greater than your previous symmetry.

Comparison of the Yang–Mills Theory and Yang Chen-Ning's Nobel Prize-Winning Work

Yang Chen-Ning: I would say that the influence was very large and long-lasting because this was a principle. There was only one class of equations, Maxwell's equations, and now it suddenly turns out that Maxwell's equations are only the simplest of a class of equations, and that class of equations now turns out to have important results for nuclear force, which is called the standard model. In fact, this is dominating the whole world of physics now, and it will continue to dominate, and it seems that there is still work to be done. No one has figured out how to do that yet, and that's what string theory is doing.

Ji Lizhen: Yeah, I think it's very important. The Standard Model is all done with Yang–Mills.

Yang Chen-Ning: You must have a lot of people doing string theory in Michigan.

Ji Lizhen: Maybe a few of them have left. I forgot the name of one of them, but then he went to Texas A&M. We have a lot of string theorists. It's hard to say what string theory looks like now. You say it's potential. People say it's all potential, but is it very clear?

Yang Chen-Ning: I think string theory is also in a bit of a difficult state right now, because it's been done for many years without any practical relevance. There's some wonderful mathematical stuff in it, but there don't seem to be any new ideas.

It's just that the days of being very enthusiastic about string theory are over.

Ji Lizhen: Over?

Yang Chen-Ning: So now string theory is also looking for a way out.

Yang Chen-Ning's Mathematical Problem

Yang Chen-Ning: I was going to ask you a question. I thought you gave me that book, the one by Felix Klein, and I took a look at it. Isn't Felix Klein's most important work on automorphic function?

Ji Lizhen: Right.

Yang Chen-Ning: Why is automorphic function important?

Ji Lizhen: Because it describes the law of a lot of sequences; the Pearson coefficient is in the form of an automorphic function. For example, if you have a sequence of numbers, a sequence of numbers that has no rule, and if you put it all together, then it becomes an automorphic form.

Yang Chen-Ning: A series of … what do you mean?

Ji Lizhen: Numbers. You say a_1, a_2 … You put it all together, nothing. But if you put it here, then it becomes a generating function.

Yang Chen-Ning: Do you mean the zeta function?

Ji Lizhen: Yeah, it's kind of like L function, zeta function, and modular form.

Yang Chen-Ning: The L function is a generalization of the zeta function.

Ji Lizhen: Yeah, a generalization.

Yang Chen-Ning: What is the relationship between the L function and the automorphic function?

Ji Lizhen: Given an automorphic form, I can construct an L function and guess roughly that basically L functions come from the automorphic form.

Yang Chen-Ning: Who first invented the field of automorphic functions?

Ji Lizhen: Poincaré and Klein.

Yang Chen-Ning: Poincaré and Klein got into a fight later.

Ji Lizhen: Yeah, because of this thing.

Yang Chen-Ning: When these were developed, were they related to the zeta function? Or not?

Ji Lizhen: No.

Yang Chen-Ning: Because an automorphic function is a generalization of an elliptic function.

Ji Lizhen: Mathematicians like Euler started learning elliptic functions, Abel and Jacobi generalized it, then Riemann and Weierstrass generalized Abel's from that direction. Later on, came Poincaré and Klein.

Yang Chen-Ning: Yes, as far as I know, elliptic functions are wonderful. They are periodic. But that group is an Abelian group, and they said let's use an SL(2, Z), which was non-Abelian. And it's not for the whole complex plane; it's for half the complex plane, and that's what they did. When they did that, it had nothing to do with the zeta function.

Ji Lizhen: It doesn't matter.

Yang Chen-Ning: The zeta function is related to that later.

Ji Lizhen: Yes, it came later.

Yang Chen-Ning: So, are they affected by the periodicity of the elliptic function?

Ji Lizhen: Yes, very much so. One of the main motivations was to prove the uniformization theorem.

Yang Chen-Ning: A uniformization theorem?

Ji Lizhen: Have you ever heard of the uniformization theorem of Riemann surfaces? Let's say that a curve $P(z, w)$ is equal to 0, and that P is a polynomial in z and w. In plane C^2, define an algebraic curve. This equation defines w as a multivalued function, not the usual single-valued function, and transforms it into a single-valued function. That is, you find a new variable and express z and w as a single-valued function through this new variable. This is called a uniformization theorem. This theorem is a very big theorem in mathematics, and Poincaré and Klein ended up in a bad relationship over it.

Yang Chen-Ning's Hypothetical Dream to Study Mathematics

Yang Chen-Ning: You know, I said a few years ago that if I were a graduate student again today, I would do mathematics.

Ji Lizhen: Because math is more interesting, right?

Difference Between Mathematics and Physics

Yang Chen-Ning: Because mathematics is different from physics. There are many directions in mathematics, and they are wonderful. Physics has ups and downs at different times. Today is not the most prosperous time.

Ji Lizhen: Yeah, but you physicists have a lot of insight that mathematicians don't have.

Yang Chen-Ning: Mathematics, it's kind of like art. It can go in all kinds of directions.

Ji Lizhen: Yes, well, thank you! I'll come back in March, and we'll see you then.

The Seventh Interview

Time: March 6, 2019
Place: Yang Chen-Ning's Villa
Interviewer: Ji Lizhen, Wang Liping, Lin Kailiang
Recorder: Wang Liping
Collator: Peng Cheng, Wang Liping, Ji Lizhen

This was the first time that we interviewed Dr. Yang Chen-Ning at his villa at Tsinghua University. The interview gave us insight into Dr. Yang's daily life. Dr. Yang was still full of energy, and the entire interview lasted more than two hours. Finally, it was time for dinner, and Weng Fan had to interrupt us. Great scientists naturally left a lot of great works, and we talked with Dr. Yang about how to reorganize and publish the valuable works and articles left by these great scientists. For example, Hua Loo-Keng's love for mathematics expressed in his articles and books was very inspiring to the younger generation, and Atiyah's outspoken comments on mathematics and mathematicians were thought-provoking. This interview also focused on the personalities and work of physicists and mathematicians, such as Chern Shiing-Shen and André Weil's work on the Gauss–Bonnet formula. Some physicists, such as Feynman, Murray Gell-Mann and Heisenberg, were surprisingly clear-cut in their attitudes to mathematics. Dr. Yang was a physicist, but he knew many mathematicians, the work they did, and the complex relationships among some of them. Many great men seemed to care a great deal about the ranking and classification of other great men in the world, and Dr. Yang talked about his thoughts on Atiyah's list of the 70 greatest mathematicians in the world. Dr. Yang also shared with us his observations on the politics of mathematics and science, especially when it comes to the supercollider. After the interview, we all expressed our admiration for Dr. Yang's superhuman memory, extraordinary learning and candid views.

Ji Lizhen: Dr. Yang, the other day I went to Shanghai and visited East China Normal University Press. We saw a painting of you there. Have you seen that painting? It's on display in a bookshop next to a restaurant. It's new.

Painting of Yang Chen-Ning at East China Normal University

Yang Chen-Ning: This one was painted last October. The thing is Zhang Shoucheng's wife graduated from East China Normal University in mathematics, so Zhang Shoucheng told them to arrange me, Zhang Shoucheng and his wife to visit East China Normal University last October. So, I visited there for two days, and I saw Zhang Dianzhou.

Do you know Tan Shengli? He's the head of their math institute now, isn't he? He probably works with Zhang Shoucheng's wife because she was a mathematics major.

East China Normal University Press's Publishing Plan for a New Book Series

Ji Lizhen: Oh, I didn't know that. I went to East China Normal University for a project at Shanghai Mathematics Publishing Center of East China Normal University Press. They needed my help with the publication of some math books. I suggested publishing a set of *Master Forums*. I meant good students need to learn from masters and read the original words and works of the masters. Upon being processed by others, a lot of good stuff in their original works has been removed, so I plan to have a few of these books published. We interviewed you many times and recorded each interaction. After transcribing and sorting out all audio and voice memos into text, plus this one, we will publish a book. This is the first book in our plan. The second book is about an interview with you in the Simons Center on your mathematics and physics dictionary, your influence on Singer and Atiyah concerning their gauge field theory, as well as on Simon Donaldson. I want to sort out part of the interview and add some contents from your lecture *Beauty and Physics* at Tsinghua University, in which you compare Dirac and Heisenberg both in terms of their scholarship and their socialization. You spoke very well. We could put your original words together and compile a book. Lin Kailiang translated some of Dyson's stuff a while back, which could also be used to publish a book. I also plan to compile a book about Hua Loo-Keng. Hua Loo-Keng wrote a lot of articles to teach people how to read. For example, his first famous article is about the solution of the fifth-degree equation, and while most people know about this article, few people have read it, so I want to put these articles directly into a new compilation. Yes, we have begun to prepare to do these, and then expand further. We will name this set of books *Master Forums*. The press president quite supports my plan. Now I'd like to hear what you think about this idea first.

Yang Chen-Ning: I think it depends on which articles you choose, and it doesn't have to be Chinese scholars, does it?

Ji Lizhen: Not only Chinese scholars but also foreign scholars. For example, after finishing your three books, I will do Atiyah's. He has written a lot of articles and has a lot of insights. I'll ask for your opinion on Atiyah later. We also want to publish a book by Poincaré. Poincaré is a very thoughtful person. But it's a little difficult to publish a book by him, because many of Poincaré's writing are in French.

Yang Chen-Ning: Poincaré has some very famous articles, but they are mostly philosophical ones, and Jacques Salomon Hadamard has some, too. As for Poincaré and Hadamard, have their writings ever been translated?

Ji Lizhen: Many have been translated into English. Some of Poincaré's books and some of Hadamard's books should be translated into Chinese. But I would like to add a little bit of mathematics here besides his philosophical stuff. Poincaré had a lot of ideas about mathematics, and I don't think anyone has had more influence in the last 100 years. I also wanted to do a book on Einstein, and since Einstein really liked to write, he wrote a lot. There are 16 volumes of his complete works, but few individuals would have the patience to read them. I would like to find someone who will read them all and choose a few articles. It would be good to be able to compile those that are of interest to the average person, because I've never even read the complete works of Einstein. What do you think of this idea?

Yang Chen-Ning: There are very few who have attempted to popularize Einstein's science. I think the easiest way for layman to learn more about Einstein is through the stories he wrote about his friends. His friends were gone, and so he wrote stories about them, and he put them all together into a book, like Madame Curie, or like Leopold Infeld, Leopold Infeld was his friend, and perhaps you've read these.[59]

Ji Lizhen: The book doesn't ring a bell with me.

---

[59]Einstein and Infeld wrote the famous book *The Evolution of Physics: From Early Concepts to Relativity and Quanta.*

Yang Chen-Ning: There is a small book that collected many of Einstein's articles or talks, for example, his views on the future and development of physics, as well as those I have just mentioned. This little book has become a pocketbook. I'll email you the name of it later.

Hua Loo-Keng's Enthusiasm for Mathematics

Ji Lizhen: I also think that sometimes what is very important to students is not what these masters did, but how they did academic work, how they dealt with problems. I think it's also very important to learn this kind of method, this kind of vision, because sometimes knowledge may be replaced by new knowledge with the change of time, but why they succeed, why they are willing to do this kind of work, which I think is a great thing. Take Hua Loo-Keng as an example. Among all the mathematicians I know, I think the one who likes mathematics most is Hua Loo-Keng. He really likes mathematics, and he is willing to share it with you. I don't have this feeling that Chern Shiing-Shen would do such things. Dr. Hua Loo-Keng told people how to study, how to do research, how to work, how to act in the classroom, and what kind of self-study they should do. Yet he was willing to compile a lot of textbooks, such as *Higher Mathematics*, from his work with college students, and other popular science topics from middle school students, so he later wrote the book *From the Unit Circle* to tell students how to start to do their own research. I think this kind of spirit is very valuable. Also, because he had a wide range of math skills, he often combined number theory, geometry, exponential forms, PDE (partial differential equations), and a lot of things together, which are particularly important for students. I want to select something mathematical from Poincaré as he is amazing. You see, he started with differential equations, then he did group theory, and he did a lot of other things. Why did he do that? I think because a lot of students learn things of narrow range, especially in China, and so do teachers. It's very important for both students and teachers to learn Poincaré's attitude and ability. It seems that he has written three books, books on the motion of celestial bodies' motion, because he became a professor in astronomy. He had to teach, and he had to teach well, so he ended up writing books. I don't think the ordinary professor has that gumption, and certainly not the average student. Therefore, I told the president of East China Normal University Press that none of the books are easy, but at least people can feel the spirit of learning after opening the books. Sometimes, the contents are not as important as how the master dealt with the problems. I think so anyway. And then I think if you have time, we can show you some of the topics first, because I've made a preliminary selection: I've chosen 16 people. Later, the publishing company said that it was very troublesome to solve the copyright problems, and we need to solve the copyright problems when we translate things from abroad, too. So, I want to do three books first, and because you have agreed to authorize your interview and Hua Loo-Keng's copyright is easier to solve. Maybe we'll do Serre's later. His subjects are more mathematical. The next one is by Atiyah, and there are books by Dyson, too.

Atiyah's Comment on Mathematics and Mathematicians' Outspokenness

Yang Chen-Ning: I haven't read Atiyah's science talk. Did he make any?

Ji Lizhen: It's not a science talk. Atiyah just told people what good math was and how to learn it. Atiyah commented on a lot of things, such as how math has developed, and sometimes he elaborated an area of math.

Lin Kailiang: He wrote a very famous article called *Mathematics in the 20th Century*.

Yang Chen-Ning: Did Atiyah write it? When did he write it?

Lin Kailiang: In 2000.

Yang Chen-Ning: I didn't read that one.

Ji Lizhen: Because Atiyah was like this, Atiyah had the audacity to stand up and say what was good and what was not, but I don't think an ordinary mathematician is always willing to stand up and give such opinions. I think that's what it looks like. It's not exactly popular science, though some books are popular science; some books are an introduction to subjects, and some are about how to do studies. For example, Atiyah later wrote an article with his students.

Yang Chen-Ning: Is the article you were just talking about found in *Atiyah Collected Works*?

Lin Kailiang: Yes.

Yang Chen-Ning: Did he basically put all his articles on mathematics and related to mathematics in it?

Lin Kailiang: Pretty much, I think so, yeah.

Ji Lizhen: That's what he wanted. It's a complete collection, not an anthology. It's all there. That's why I think we could publish a book with stuff written by Atiyah. I can't remember all the names of the 16 people I just told you about. I only have a general idea.

Lin Kailiang: Did Weyl have any?

Yang Chen-Ning: Weyl.

Ji Lizhen: Weyl, yes, I think I should include him. There's also Gorō Shimura, a Japanese mathematician. You may know him. I heard that he wrote a lot of books in Japan, and he was very bold. Why is he bold? Because the whole development of mathematics in Japan was led by Takagi Teiji, so Gorō Shimura was very bold to write, and he dared to criticize Takagi Teiji.

Yang Chen-Ning: Did he know André Weil very well?

Ji Lizhen: André Weil was at Princeton with him, but Gorō Shimura was a generation older.

Yang Chen-Ning: There's a famous conjecture.

Ji Lizhen: Yes, that's him. The Taniyama-Shimura-Weil conjecture was proposed by those three guys.

Yang Chen-Ning: He's still around?

Ji Lizhen: Yes, he is.[60] He's still at Princeton, but he's very old.

Yang Chen-Ning: He's at Princeton? You said he was at Princeton for many years?

Ji Lizhen: Yeah, for decades. He was at Princeton for many years. Last time I found a useful pamphlet in the library of the University of Tokyo, it was about Gorō Shimura's learning, in contrast to Confucius' theory. If anyone could find this pamphlet, they should print it again. But I'd like to send you the list first.

Yang Chen-Ning: I think so. The plan you just mentioned can have a wide range, so it depends on how you choose.

Ji Lizhen: Yes, it also depends on our first printings. If we do well with the first volumes, it's all good. We did *Mathematics Overview* with Higher Education Press, and it was relatively successful early on.

Wang Liping: More than 20 books have been, which have received good response in the market.

Ji Lizhen: Yes, but I didn't choose stuff by big names only in *Mathematics Overview*. There are stuff written by other people, too. If we're doing it in Shanghai, I think we'll just publish the original writings, and add some notes, so basically, we'll just publish the original. Why shall we

---

[60]Gorō Shimura died on May 3, 2019.

do this? It's because I've read a lot of domestic books about Hua Loo-Keng, but they were rewritten, and what Hua Loo-Keng himself wrote was not the same, so the books were not original to their source.

Chern Shiing-Shen and André Weil's Work on the Gauss–Bonnet Formula

Yang Chen-Ning: I only know Chern's famous article of "A simple intrinsic proof of the Gauss–Bonnet formula for Closed Riemann manifolds." After that article came out, André Weil wrote a review, which Chern said was very important. This review was closely related to the direction of Chern's subsequent development.

Ji Lizhen: Really? Let me have a look.

Yang Chen-Ning: There's a magazine called *Mathematical Review*.

Ji Lizhen: Yeah, I know about it. Chern's Intrinsic Proof of the Gauss–Bonnet formula is a very famous result, and it was very much influenced by an article by André Weil. Take this example. I think it's important to read some of the people's articles or their comments. I want to encourage students to read this kind of stuff. I wonder if I could consult with you on this set of books.

Yang Chen-Ning: No need. I can actually be a consultant, but don't put my name on it.

Yang Chen-Ning's Neighbors at the Institute for Advanced Study and Their Work

Ji Lizhen: I really think your interview book is very important.

Yang Chen-Ning: This one is special because it covers a lot of things, and it's hard to say that there are fixed issues, including how Uhlenbeck got into the University of Michigan, and it's the first time for me to know that you used to know Borel very well. This Borel had nothing to do with the Borel of the 19th century,[61] right?

Ji Lizhen: No, many people have asked him the same question many times, and me too.

Yang Chen-Ning: Although Borel was our neighbor, we rarely talked to each other. I don't know if he had any dealings with Dyson, because we were all close neighbors. My house, Dyson's house, and Borel's house were all close to each other, and Harish-Chandra's house was a little further away.

The house of Harish-Chandra used to be John Archibald Wheeler's. I remember Wheeler built this house in the 1930s and 1940s on the grounds of the Institute for Advanced Study, then Wheeler moved somewhere else, and Lee Tsung-Dao lived in it. Lee Tsung-Dao lived there when he was a professor for two years at the Institute for Advanced Study from 1960 to 1962. After Lee Tsung-Dao left in 1962, Harish-Chandra came, and then Harish-Chandra lived there.

Ji Lizhen: Yeah, it's amazing. It's not very impressive if people go there and they don't know who lived there before, but it's different when we do.

Yang Chen-Ning: I have a question for you, too. I don't remember who I asked last time. Maybe I asked you what was going on at Langlands. The answer was that he was working on the representation theory of automorphic forms. The automorphic forms are closely related to this Langlands discrete group, right? Are all automorphic forms dissimilar discretions?

Ji Lizhen: That's right.

Yang Chen-Ning: So then, it stands for representation theory, which is actually a representation of noncompact discrete groups.

---

[61]Borel in the 19th century is known for his work on measure theory, and the present Borel is known for Lee theory and many related applications.

Ji Lizhen: No. It's like, let's say the trigonometric functions sin and cos, because they're periodic, for this Z, and then the higher dimension is Z^2; suppose the two-dimensional ones are lattice, the three-dimensional ones are lattice, and then this group is the Abelian group, so let's take a non-Abelian unbounded group, like SL(2, Z), if now …

Yang Chen-Ning: It's non-Abelian, and it's not finite.

Ji Lizhen: Yeah.

Yang Chen-Ning: Wasn't Harish-Chandra's important work on non-compact groups?

Ji Lizhen: He dealt with Lie groups. It's not a discrete group. A Lie group has a discrete group in it—for example, the real number R has a Z in it; it's on the Z axis up here that we see periodic functions. Let's say R^2 has a Z^2 in it; the group SL(2, R) is a Lie group, where you replace the coefficient R by Z. Then it's an infinite discrete group.

Einstein and Other Physicists' Views on Mathematics

Yang Chen-Ning: Now I suddenly have an idea. I think there is a very interesting topic, which is the relationship between mathematics and physics. Regarding this relationship, different mathematicians and physicists have changed their attitudes at different times. In his late year, Einstein said that when he was young, he did not understand the usefulness of mathematics. He did not go into mathematics because he could not grasp what was important and what was not important in mathematics. Later, I guess when he moved from special relativity to general relativity and got to know Minkowski's work, he changed his mind. So, by the end of his life, he was saying the opposite of what he had said when he was young, that he thought mathematics was very important.

Ji Lizhen: Mathematics is very important, yes.

Yang Chen-Ning: Another very interesting one who's a little bit similar is Heisenberg. You can tell a lot from what Heisenberg said in his early years—that he was contemptuous of mathematics. But he also said something later in life that makes you think that he's changed a little bit, not as much as Einstein, but obviously changed. It's only half the difference. Or like Feynman and Gell-Mann, and they don't think mathematics is useful, either. Feynman repeatedly said that the mathematics we need can be created by ourselves, which reflected his own experience, but he generalized it in a way that I think would be incorrect. He was reflecting on his own feelings because the path integral, for example, was a basic mathematical concept, and it was a mathematical concept that he used, which was very useful, but it has not been established by mathematicians even up to now.

Ji Lizhen: No, it's not strictly established.

Yang Chen-Ning: He didn't believe in math that came out of education. He only believed in math that emerged from guessing.

Ji Lizhen: Yeah, I read Landau's biography translated into Chinese not long ago.

Yang Chen-Ning: I heard that there was a Landau biography. Who wrote it?

Ji Lizhen: It's by his niece.

Lin Kailiang: Last time, I brought you a copy. You had a quick look and gave it back to me.

Yang Chen-Ning: What's its name?

Wang Liping: *Landau's Biography*.

Yang Chen-Ning: The author is not Kalashnikov?

Ji Lizhen: No.

Yang Chen-Ning: That's his student, isn't it?

Ji Lizhen: No, it was written by his niece, his wife's niece.

Yang Chen-Ning: His niece. Not a physicist?

Ji Lizhen: She's not a physicist.

Yang Chen-Ning: I think I have read it.

Wang Liping: I gave it to you last time.

Yang Chen-Ning: I remember after you showed it to me. I read it casually because I thought the author was not a physicist or a mathematician, so I didn't read it much. To me, it was just something unimportant.

Ji Lizhen: There's a lot in it about what kind of math Landau considered to be important.

Yang Chen-Ning: I think there's a book where you can get an in-depth understanding of Landau's personality. It's a book written by his students. You know it, right?

Ji Lizhen: It's translated into English, right?

Yang Chen-Ning: (The author) is called Kalashnikov. He's a student, maybe a postdoc. After he died, several of those who were very close to Landau each wrote an essay and turned it into a book. It was a very interesting book because it described how Landau taught his students, so it gave a good account of the Landau school atmosphere.

Ji Lizhen: I'll check it out. It's in English, right?

Yang Chen-Ning: I read the book in English. I think it was translated.

Ji Lizhen: It was translated from Russian.

Yang Chen-Ning: You know the book, right?

Ji Lizhen: Yeah, let me find it.

Lin Kailiang: I remember, when you were at Tsinghua University, Chern Shiing-Shen said something like this. Right here, he said that when he was teaching in Chicago, there was a gentleman who thought that physics students should not study too much mathematics. Who was this man?

Yang Chen-Ning: Chern Shiing-Shen went to Chicago around 1950 and stayed there for a decade. During that period, Fermi wouldn't have said that, neither would Teller. I don't know who Chern Shiing-Shen was talking about, but he could have been talking about Gell-Mann. Gell-Mann is still around; he's 90 years old, and he made a very important contribution to high-energy physics. He probably came to Chicago in 1952 as an assistant professor, and at that time he was the most famous young physicist of my generation in Chicago, so Chern Shiing-Shen must have known him, and he might have said that. But he didn't say it as thoroughly as Feynman did. In other words, he didn't believe it as much as Feynman did, and Gell-Mann was only half-joking.

Lin Kailiang: I remember because we put together this version, and I remember at the beginning of the Chinese Taiwan version, it mentioned Hami, and he said there is a Dr. Hami at the University of Chicago. Is there such a person?

Yang Chen-Ning: Just let me think about it. I wonder who that is.

Lin Kailiang: Maybe it wasn't well recorded at the time.

Yang Chen-Ning: At that time, unless he was a math student, there was no such person in the physics department. In the 1950s, his name or family name was Hami, but I don't know who he was, and Chern Shiing-Shen never told me that, either.

Yang Chen-Ning's Impression of Atiyah, Serre and Other Mathematicians

Ji Lizhen: I read your lecture on *Beauty and Physics* at Tsinghua University. It was very interesting and enlightening. You compared Dirac and Heisenberg in terms of their personalities and their learning. We're going to talk about some of the mathematicians you know and look at their personalities and their learning. I want to start with Atiyah because Atiyah recently died, and there was a lot of buzz about his recent announcement that he had solved the Riemann conjecture. How about your impressions of Atiyah? For example, how about his work with people or his learning.

Yang Chen-Ning: I can't comment on the greatness of Atiyah's mathematics because I haven't studied much (mathematics), but I have heard, especially from Chern Shiing-Shen very early on, that he was extremely important. He combined analysis, algebra, geometry and topology. I had some contact with him. When he was at the Institute for Advanced Studies in the 1950s, I didn't know him, and I hadn't heard that he was a very important physicist. Serre, for example, was a little bit older than Atiyah, but soon I heard that Serre was very famous. It was not that way with Atiyah. So, in the 1950s, until I left Princeton and went to Stony Brook, I had no idea that Atiyah was an important mathematician. The first time I heard about it was when I heard it from Chern Shiing-Shen.

Later, the normative field theory was developed after Wu Dajun and I wrote that article. Remember my dictionary? I remember the dictionary very well, and many other people remember it very well, like Singer. Singer and I went to graduate school together in Chicago. He was also a physics major, but he switched his major to mathematics halfway through, so his Ph.D. was in the mathematics department. I knew him, but didn't know him well. He came to visit Stony Brook in the 1970s, not visiting me, and I don't know who he was visiting in the math department, because there was a lot of stuff going on in the math department then. He came to my office, and I gave him a preprint copy with this dictionary on it. Later, he said he was taking it with him to Oxford. It turned out that the canonical field theory that we were talking about was a fiber bundle, and one of the simple mathematical problems in the fiber bundle was called instanton.[62] Instanton had already been written by three or four famous physicists. The paper was quite well known, which made it known that some physicists had made a wonderful mathematics achievement, and they called it fiber bundle. One simple example of a fiber bundle was a monopole. The one for U(1) was too simple and mathematicians were only interested in the more complicated SU(2). The simplest SU(2) is a four-dimensional mathematics problem, i.e., instanton. Using the index theorem, people can know how many parameters the solution has. This makes mathematics … Even though mathematicians already knew more or less what physicists were doing in those days that had something to do with mathematics, it was still all vague. In fact, a lot of people … If you read the paper written by Wu Dajun and me, we have a very long footnote. The reason is that before us, mathematicians and physicists were more or less aware of the close relationship between electromagnetism and fiber bundle theory, but they had not clearly figured it out. So, people wrote articles one after another, with both mathematicians and physicists involved. When we were writing that article, we said we would have to be careful; otherwise, these people were going to say we stole their ideas. So, we wrote down as many names as we could, and some of them were famous mathematicians, like a famous French mathematician who did differential geometry, whose name starts with the letter L.

Ji Lizhen: Because he worked with spinors. I know his name was André Lichnerowicz.

Yang Chen-Ning: We just put them all on the list; that's why. I don't think they know what our contribution was because they are two different concepts, and our important contribution was to make the dictionary so that the two concepts would be seen as the same thing, and that they are one for one. And I think the one that's most interesting

---

[62]Instanton is the solution to Einstein's equations and other field equations in which time and space coordinates cannot be distinguished.

to mathematicians is the Dirac monopole, which is a nontrivial fiber bundle that goes from a trivial fiber bundle to a nontrivial fiber bundle, and this becomes mathematically very meaningful. This, I think many people, including Lu Qikeng …

Lu Qikeng and Yang Chen-Ning

Lin Kailiang: So, a teacher in our school was one of his students. That teacher knew I was talking to you, so he mentioned it to me.

Yang Chen-Ning: Here's the thing. I gave a speech (in China) in 1972. It was an important speech. Why? Because I gave a number of lectures in 1972, two of which were now recorded, one on the theory of normative fields and the other on solvable statistical mechanics. Both of these fields were at an important stage of development, of which China was completely unaware.

Ji Lizhen: China was very closed off in the 1970s.

Yang Chen-Ning: It was not just about being closed off, but about a hierarchical model. They were fascinated by the hierarchical model. Everything was hierarchical model, and anyone who went out of the field would be excluded. So, I told China about the importance of these two fields, and as a result (some people) immediately became interested in gauge field theory, in addition to Gu Chaohao, who was in mathematics, and Li Huazhong, who worked at Sun Yat-sen University … Because there was a group of people in Beijing, they suddenly found that the standard field theory in international literature at that time had become very important, and they were a little afraid, so they would not allow people to do this thing, and they crowded out Li Huazhong. Finally, in the late 1970s, they couldn't stop it. They only allowed young people to do it later. However, Lu Qikeng probably understood something from my speech and wrote some articles, which he himself said he sent to me. I read them, but I did not understand them because he used the language of mathematicians. I read it and did not understand his meaning. There were many such articles. He later thought that Wu Dajun did not quote his article when he wrote with me in 1975. In fact, I had no idea what he was doing, so he was always unhappy. The Frenchman's name starts with an L. What's his name?

Ji Lizhen: André Lichnerowicz.

Yang Chen-Ning: He wasn't happy with me afterwards, either. He was a very famous mathematician.

Ji Lizhen: Yeah, because he proposed an elimination theorem, a very important equation showing when certain spinors disappear, so André Lichnerowicz is very important. Atiyah, of course, is a very great mathematician, and I think one of the other important things, more than any other mathematician, is that he's also a public figure. What do you think about that? Because he actually came out and gave a talk, and he got a lot of media attention, which ordinary mathematicians don't get in general.

Yang Chen-Ning: My impression? My impression is not from his work, because 1 did not get his works in mathematics, but I think I appreciate what Weyl said. When Weyl said that regarding mathematics, he liked mathematics with a broad vision, I suspect that Atiyah, André Weil and Weyl all had the experience and ability to move in that direction. That's my general impression.

Ji Lizhen: I've also heard that Atiyah's most famous work is the index theorem, along with Singer's.

Atiyah and Singer

Yang Chen-Ning: I remember it was Atiyah who worked on the theorem, but when it came to differential geometry, he didn't understand something, so he (collaborated) with Singer. My impression was that basically Atiyah had the idea, but it needed to be solidified in terms of differentiation, and this concretization was developed by Singer for him. Chern Shiing-Shen was quick to tell me how important the article was.

Ji Lizhen: Did Chern Shiing-Shen see it that way?

Yang Chen-Ning: I remember he said it was the most important mathematical article in recent years.

Ji Lizhen: I think Atiyah was very broad-visioned because he was able to apply the index theorem in many ways. Later, Witten came out, which made Atiyah admire even more the relationship between mathematics and physics.

Yang Chen-Ning: After Witten came out, Atiyah appreciated it very much. I guess that Atiyah nominated him for the Fields Medal.

Ji Lizhen: Yeah, he pushed for it. Good.

Yang Chen-Ning: So, if you want to ask me my impression of Witten, I don't understand what Witten is doing. It seems that Witten is a very talented mathematician. He had some ideas about many mathematical things before they were specific, which is what other mathematicians admire very much. But what he had done had not yet been translated into physics.

Lin Kailiang: No application?

Yang Chen-Ning: No, he made some contributions to the theorem of positive mass.

Ji Lizhen: There is a new proof. Witten has given a very good proof.

Yang Chen-Ning: Of course, it has something to do with physics. It's not an important conclusion in physics, though.

Atiyah's List of 70 Great Mathematicians

Ji Lizhen: Yeah, it's more mathematical physics. There's another thing Atiyah did that I don't know if you're aware of. Atiyah and his wife chose 70 mathematicians in history, the ones who had influenced him the most.

Yang Chen-Ning: Did you also write an article on this?

Ji Lizhen: No, he has a website. (URL: http://www.maths.ed.ac.uk/~v1ranick/atiyahpg.pdf)

Lin Kailiang: I sent you a photo album, a small photo album, with some mathematicians in the photos.

Yang Chen-Ning: 70 mathematicians?

Lin Kailiang: Yes, 70 mathematicians who are related to him.

Ji Lizhen: The ones he thought were important.

Yang Chen-Ning: Could you email these people's names to me? He has 70 people. How many of the 70 people were from the 20th century?

Ji Lizhen: Quite a few.

Lin Kailiang: Most of them were from the 20th century.

Yang Chen-Ning: Most of them were from the 20th century?

Lin Kailiang: People who are connected to the couple.

Ji Lizhen: Because he had a lot of students …

Yang Chen-Ning: Was his wife a math major, too?

Ji Lizhen: Yes, she teaches math, but doesn't conduct research.

Yang Chen-Ning: I haven't met his wife.

Ji Lizhen: He and his wife were classmates in graduate school. She also studied mathematics there. I think it's also important for young people to know who is more important if someone is willing to stand up and speak out.

Yang Chen-Ning: That part depended on how Atiyah told it. Whether he said that these were the most important mathematicians of all time, the most important mathematicians in his opinion, or whether he said that these were the mathematicians that he believed to have the most influence on his work. That's different.

Ji Lizhen: He said that, and that they had an impact on his work, but people tend to interpret it as … Because people naturally react, and Atiyah, being a famous mathematician, when you put names like that out there, people automatically assume that you think highly of them.

Yang Chen-Ning: I think the list is very interesting, especially in the 19th and 20th centuries. What kinds of talent are on the list? Is there someone like Felix Klein?

Ji Lizhen: I don't remember Felix Klein. Maybe not.

Lin Kailiang: Maybe not.

Yang Chen-Ning: Is Lie already on it? Everything is related to Lie groups.

Lin Kailiang: Yes, I think so.

Ji Lizhen: One interesting thing about Atiyah is that when Atiyah was a graduate student, André Weil was passing by Cambridge and tried to persuade him not to study mathematics.

André Weil's Attempt to Dissuade Atiyah

Yang Chen-Ning: Did André Weil talk Atiyah out of it?

Ji Lizhen: He told Atiyah not to study mathematics.

Yang Chen-Ning: Don't study mathematics?

Ji Lizhen: That's right.

Yang Chen-Ning: Why?

Ji Lizhen: André Weil probably felt that Atiyah was just not good enough, that he would never be good enough.

Yang Chen-Ning: That was when Atiyah was very young.

Ji Lizhen: Yeah. I heard that when Atiyah graduated, he wrote a senior thesis and sent it to André Weil. When André Weil got it, he took out a piece of paper and wrote heavily on it. When the paper was taken away, there was no writing on it, but you could see very clearly what he had written. André Weil had said, "This is junk."

Yang Chen-Ning: Did they compete on any issues?

Ji Lizhen: I don't think so. André Weil was probably a generation older than Atiyah, and they did different things. All I can say is that maybe André Weil didn't think Atiyah was capable enough.

Yang Chen-Ning: Was Atiyah's speech ambitious or not for a mathematician?

Ji Lizhen: He's not very ambitious.

Yang Chen-Ning: I don't think he was that ambitious, either.

Ji Lizhen: He's not very ambitious.

Yang Chen-Ning: André Weil is a tougher guy who likes to criticize people. I remember there was a mathematician at Princeton who wrote *The Biography of Number Theory*. Do you know about that?

Ji Lizhen: *Number Theory*? Oh, Fermat, *The Biography of Fermat*.

Yang Chen-Ning: It turned out André Weil wrote a review.

Ji Lizhen: Yeah, I read that review.

Yang Chen-Ning: In the review, he said that a person who wanted to write a biography of number theory must know mathematics and ancient Chinese characters. The third condition, I've forgotten, but he said that this gentleman could not meet any of the three conditions. And that's very sad because that guy probably didn't have tenure at that point, and he was done for after this. Do you know who this guy is?[63]

Ji Lizhen: Yeah, I read the review, and one of the things I saw was that André Weil was very proud, then he came up with a lot of ways to say it, and it made you feel that he was proud.

Yang Chen-Ning: Did I tell you the story last time that he wasn't happy with me? Then he took his dissatisfaction to ... There was a physicist at the University of Chicago who was ten years younger than me named ... whatever, I can find that out later. Anyway, he wrote a book about his experiences with me in the Soviet Union, and so on, but there's also a paragraph in the middle about André Weil, which shows that André Weil was very unhappy with me. I'll tell you why, and then it was over the appointment of a historian at the Institute for

---

[63]The title of this book is *The Mathematical Career of Pierre de Fermat*, 1601–1665. The author is Michael Sean Mahoney. Mahoney received tenure at Princeton University. André Weil seems to have had an academic dispute with Mahony's mentor, also at Princeton, a very famous historian of science, Thomas Kuhn. And André Weil didn't have much regard for historians of science and mathematics who didn't know the subject very well.

Advanced Study. I remember there was a very loud argument. I said at a conference that I came from an ancient culture that had a general view of the world, not one directional. What I meant by my remarks was that the Institute of Advanced Study at that time had said that this man was a great man and that they were very eager for him to join them, which I don't think is in line with China's need to look at things in a broader context. This made André Weil very unhappy, and he felt that he, too, was from an ancient culture.

Ji Lizhen: His culture is probably a little younger than ours.

Yang Chen-Ning: I didn't know about it, but he went to talk to a young physicist at the University of Chicago who wrote about it. I'll find the book next time.

Lin Kailiang: I think you've sent me something about it. It was called a *Passion of Discovery*, wasn't it?

Yang Chen-Ning: Yeah, that author seems to have died recently. I'm posititve about this. André Weil was one of the most important mathematicians of the second half of the 20th century. He was also a very critical person. But his contributions, I think, were multi faceted and will always be recognized.

Ji Lizhen: He left, and at 60, he stopped doing mathematics. He did the history of mathematics, which is also very important. He seemed to be the only person who really knew math well enough to do the history of mathematics. The whole André Weil thing had an impact on Atiyah because there was later a book published by Princeton called the *Princeton Companion of Mathematics*,[64] and there was advice from some famous mathematicians to their students in a passage at the end of it, including an essay by Atiyah. Atiyah said that if a student had not experienced a period of time when he had doubts about his ability, he was not a very normal student. If a student had a period of doubt

---

[64]The Chinese title is 普林斯顿数学指南, and the English version is a large-scale work edited by G. T. Gowers, a Fields Medal winner, and co-written by 133 famous mathematicians. The Chinese translator is Mr. Qi Minyou.

about whether he could do this thing, maybe he still had a chance. After I saw it, I felt quite impressed.

Yang Chen-Ning: The most contact I had with Atiyah took place when I asked him to be the director of the mathematics review committee for the Shaw Prize. He did that for a good number of years, and then he said that because of it, he had to fly to Hong Kong SAR twice a year, and he thought it was too much, so he stopped.

Ji Lizhen: I just thought of this. In *Beauty and Physics*, you compared the personalities and academics of Dirac and Heisenberg.

Yang Chen-Ning: The difference between Heisenberg and Dirac is quite meaningful. You could say Heisenberg and Dirac were people who thought in different directions, totally different. Their intuition and the way they represented themselves were not the same. I can also tell you a story that I probably didn't tell you last time. I was 60, and Dirac was 80. Dirac had a conference at Trieste (International Centre for Theoretical Physics), and Dirac was there, so I had a lot of contact with him. In the middle of it, I asked him a question, and I said, "Where do you think the next breakthrough in fundamental physics is going to be in mathematics? Is it algebra, topology, analysis or geometry?" I could totally predict that he would be talking about algebra, and why, because his most famous paper is the Dirac equation, which uses anticommutation of images, anticommutation of matrices, matrices and so on. But what he gave me was something I didn't expect at all, and he said analysis, and I didn't know about any of his work where he used a lot of analysis. I still don't know.

Ji Lizhen: Does it have something to do with the Schrödinger equation? The Schrödinger equation also requires a lot of analysis.

Dirac's Most Important Job

Yang Chen-Ning: The other thing that made me a little surprised by his answer, and this will surprise all physicists, is that I asked him, "What do you think is your most important work in physics?"

Ji Lizhen: Yeah, and what did he say?

Yang Chen-Ning: I think 90 percent of theoretical physicists would say it's the Dirac equation, but he said no. He said it should be transformation theory, and you know what he said. I thought about it, and I understood.

Ji Lizhen: And what did he contribute to the theory of transformation? Which group theory is it?

Yang Chen-Ning: This means that the Heisenberg equation is to be studied using Hilbert space theory.

Ji Lizhen: Which group theory does he use?

Yang Chen-Ning: I don't think he was aiming at that because there is a distance between a matrix and a Hilbert space.

Ji Lizhen: Yes, one is finite-dimensional, and the other is infinite-dimensional.

Yang Chen-Ning: The matrix is discrete, and the Hilbert space becomes particularly large. But I think, and this is my guess, from my understanding of Dirac, who understood both matrix theory and this Hilbert space, but didn't understand how mathematicians built the relationship between the two, he took this matrix theory that he knew and applied it to this infinite dimension. So, both his noncommutative variable and his mathematical notation were that you could solve from a matrix or that you could solve from a Hilbert space. So, as soon as he saw Heisenberg's first paper, and you know this is a famous story, he thought that the noncommutative variable was related to the Poisson bracket. He had this idea, but he couldn't remember what the Poisson brackets were, so he waited, and because it was Sunday, the library was closed. And on Monday morning, he rushed to the library and looked it up. He found out what this Poisson bracket he had forgotten was. He knew it had something to do with Heisenberg's theory of noncommutative variable, and he soon wrote a paper on what he referred to as transformation theory. I think you can actually understand … Because I think from what he said, you could say that the mathematical concept of Hilbert space is very fundamental to Heisenberg's theory, and I was the first to know about him. And then he wrote a book, and that book was Dirac's quantum mechanics. If you look at this book, his contribution was not that he fully understood the relationship and difference between matrix theory and Hilbert spaces, which I don't think he ever truly understood. But he did, from what he knew about matrix theory, guess about Hilbert spaces, and then he applied that to quantum mechanics, which he did first.

Ji Lizhen: So, that's why he thought the analysis was important, because there was some analysis involved in Hilbert spaces.

Yang Chen-Ning: Why is that?

Ji Lizhen: Does that explain why Dirac thought the analysis was so important? Because Hilbert space theory requires a certain amount of analysis, shall we say?

Yang Chen-Ning: He didn't think he knew Hilbert spaces; he thought he didn't know Hilbert spaces, but he knew that it should be applied to Heisenberg's work, so he appreciated that. No one's actually written about this, but this is my theory now, and you saw him say that transformation theory is the most important. So, I think there were two things that surprised me at that time: one was what I just said, and he thought it was his most important work, and the other was that he thought the future of physics would be analytical and not algebraic. Both of these things would surprise you physicists.

Lin Kailiang: Did he also say in his later years that he thought noncommutation was important in his early years?

Yang Chen-Ning: What did he think? When I said this at the beginning, he said that in the 1930s, he thought the most important thing about quantum mechanics was non-commutative variable. Now he doesn't think so. Now, he thinks the most important thing is that quantum mechanics introduces a phase. This is what I quoted him saying.

Ji Lizhen: Now I have a question. What is the difference between Chern Shiing-Shen's and Hua Loo-Keng's working style?

Yang Chen-Ning: In my opinion, Chern and Hua are very different. Hua has a lot of new ideas, and he builds them very quickly. Chern Shiing-Shen seemed to be quite single-minded. Chern Shiing-Shen really understood Elie Cartan's theory of differential forms, and then Chern Shiing-Shen took this method to the extreme, which influenced the whole differential geometry. And in particular, he used it to get the generalized Gauss–Bonnet theorem, which I think is one of the most important works of the 20th century. So, I'm not surprised to see Atiyah include Chern Shiing-Shen in his list of dozens of people. Then Chern came up with the Chern class with the addition of topology. I wrote an article with the quantum number of particles probably being the most important contribution. Why are there so many quantum numbers in the world? Where do they come from? André Weil's position is not a conjecture. It has something to do with the Chern class. You need to ask him. I think his guess is probably right, but I haven't found it out yet. Until now, none of the numbers in this category can match up, so I keep saying that if they match up in the future, of course, Chern Shiing-Shen will become a Bodhisattva.

Ji Lizhen: That's amazing.

Yang Chen-Ning: It's not just arhat. I think Chern Shiing-Shen knows about it himself. That's why he said that arhat talked to Bodhisattvas.

Lee Tsung-Dao's CUSPEA and the Chern Shiing-Shen Program

Yang Chen-Ning: Here's the thing. At that time, Lee Tsung-Dao started a CUSPEA (China–United States Physics Education Association), you know?

CUSPEA was so successful that about 1,000 Chinese students came to the United States. I don't know whether it was Chern Shiing-Shen's idea or Lee Tsung-Dao's idea, but Chern Shiing-Shen said he would make a CUSMEA, changing P into M. CUSPEA means that the China–United States Physics Education Association.[65] It changed "P" into "M," which is mathematics. So, it is Chern Shiing-Shen who did what Lee Tsung-Dao had done. However, Lee Tsung-Dao's method was different from Chern Shiing-Shen's. How did Lee Tsung-Dao do it? He and his secretary worked very hard. The first step he took was to write letters to many universities, saying that now China can send students to study overseas. China has very good students, so he appealed to the universities for accepting a certain number of graduate students with financial aid. In this way, he wrote to many universities, about 50 or 60. Many agreed and wrote to Lee Tsung-Dao regarding how to select Chinese students. Lee Tsung-Dao recruited a few scholars from these universities specializing in physics, about seven or eight of them. They formed a selection committee to screen Chinese applicants. It was very simple and successful. But Chern Shiing-Shen didn't want to do it this way. The reason was that Chern Shiing-Shen didn't like to be in charge. He was a man who didn't want to interfere, and the recruitment was too troublesome. So, he said he would leave it to the American Mathematical Society. Wasn't that natural? He asked the American Mathematical Society to organize the matter. This decision made some Chinese people dissatisfied, and the general attitude was why Chern asked foreigners to do this and why he didn't ask Chinese?

---

[65]CUSPEA is short for China–United States Physics Education Association. At the beginning of the reform and opening-up, Professor Lee Tsung-Dao, a Chinese-American Nobel Prize winner at Columbia University, initiated the CUSPEA program to select outstanding Chinese physics students to study in the United States.

From 1979 to 1989, the CUSPEA program selected 915 Chinese students to study in the United States, 30% of whom returned to China after completing their studies. Over the past 40 years, CUSPEA scholars have made many remarkable achievements in the fields of physics, finance, biology, computer, education and so on. A number of them have become world-class scientists or industry leaders; 12 of them have been elected academicians in China, Europe, the United States, Canada and other countries; and more than 100 people have won various international science and technology awards. More than 300 of them have held positions in international science and technology organizations, and more than 400 of them have been successful high-tech inventors or entrepreneurs.

In other words, why didn't he ask for Siu Yum-Tong or Hsiang Wu-chung?

Ji Lizhen: Yeah, I think Chern went to Phillip A. Griffiths.

Yang Chen-Ning: Maybe Chern Shiing-Shen asked Griffiths to do it.

Ji Lizhen: Yes, Griffiths. Why Griffiths?

A Beautiful Mistake

Yang Chen-Ning: The mistake is apparently like this. A is a mistake. B means that the person is good enough to make the mistake, so the person will say it is a beautiful mistake.

Ji Lizhen: You know, when Poincaré won the King's Prize in Sweden, he wrote a paper on three-body motion, and when he printed it all, he found an important mistake when he checked it, and then that mistake developed into the whole chaotic system, or chaos theory. That mistake was so important that it changed the whole field, so that mistake would have been a really great mistake. It was very interesting.

The Physicist's Child

Yang Chen-Ning: I forget. Did I ask you if you know any of the Chinese professors in the Physics Department of the University of Michigan? One is Yao Ruopeng, Yao York-Peng Edward.

Ji Lizhen: I don't know him.

Yang Chen-Ning: One is named Wu Qitai. You don't know any of them? All are retired, of course. Wu Qitai is a mathematical field theorist and seems to be 10 years younger than me. He was a professor at Michigan for many years, and now he's retired. He was in poor health and moved to his daughter's house in San Diego. Yao Ruopeng is about his age, a Cantonese, and still in Michigan. There is another of Japanese descent called Yukio Tomozawa. I think he's still doing work … About this Yukio Tomozawa, he had a son who was with my little daughter in kindergarten school. One year, I sent my little daughter, called Julie (Yang Youli), to school. She was born in 1961. I went to her small kindergarten at the Institute for Advanced Study. There were about 20 or 30 children in it; all the children were kids of the people who worked in the Institute for Advanced Studies. For example, (Raoul Bott), whose children were there at the same time as my oldest son (in that kindergarten).

Ji Lizhen: You said last time that several parents worked together.

Yang Chen-Ning: Yes, my eldest son is seven or eight years older than my daughter, so when it came to my daughter, I went to their kindergarten one day for my daughter's birthday. There was a birthday party, so I went there. I went to see their children, with my daughter Yang Youli, and at the same time, there was a boy who was Yukio Tomozawa's son. I filmed some movies, and we still have them at home. Yukio Tomozawa's son was very interesting; he was one of the kids who showed everybody how to lie down to sleep for a few minutes, so the children all lay down, each carrying their own blanket. But this boy was not the same as the others, and the film showed it very clearly; he was especially careful, and his blanket had no wrinkles, so everything was very detailed. We noticed this at that time, and I recently saw this again. About a few years ago, I saw Yao Repeng, so I asked about him, and I said I thought Yukio Tomozawa was not in Michigan, but he said he was. I asked him if he knew Yukio Tomozawa had a son who was the same age as Yang Youli. He said he knew about it and that he had made a fortune because the child was on Wall Street now. So, I think his richness must be associated with carefulness.

Ji Lizhen: It's interesting. A person's personality is very important. It plays a big part in what he does.

Yang Chen-Ning: I think mathematicians have different personalities, and they have different tastes in mathematics.

Ji Lizhen: It's also very different because the questions you choose are very different.

Hua Loo-Keng and Chern Shiing-Shen

Yang Chen-Ning: I think some people admire Hua more than Chern.

Ji Lizhen: That's not quite right. Not long ago, I watched two documentaries, one about Hua and the other about Chern. After watching Hua's film, I was very moved because I thought he came from such a difficult background, but later he became self-educated and loved math so much. Some of Chern Shiing-Shen's stories are new to me. Later, many people said that Hua's attitude toward study was very respectable, but it is hard to say how much of a mark his contributions to mathematics will leave on history. Chern's work has had a great impact on history. It is possible that Hua valued quantity more than quality. When Hua was at Princeton, he lost a great opportunity. Weyl was there, Siegel was there, and it might have been better for Hua's future if he had studied some advanced mathematics from the masters instead of writing a lot of articles. When Hua was at Cambridge, he didn't learn much from Hardy to then make something for himself, either. Later, I thought this kind of comment was very pertinent. Could you tell us this in your speech? It is no use teaching people how to read and learn without being specific. This is a good example because I also think that Hua's contribution to mathematics is really hard to say regarding how much influence it will have on history, although he has influenced many generations of Chinese people. What do you think of him? I think his character is probably more …

Yang Chen-Ning: I think so. When Hua came to the University of Illinois, simply looking at it from his mathematical career, his return to China had a personal impact on him. In terms of his taste and ability, if he had stayed in the United States, I think he would have made a lot more progress than what he had already done in mathematics. It can be said that the new mathematics of the second half of the 20th century began at that time. With his skills and interests, if he had stayed in the

United States, even though Illinois was not the hottest place, I think he would have been affected a little bit. So, I think from the standpoint of Hua's contribution to mathematics, his return to China at that time may have affected his development. You know, when they were in Illinois, they wrote to Hua Shun, their eldest daughter, asking her to come to Illinois to find them. His younger children went there, but Hua Shun wrote back, saying that she would not go to the United States. I think Hua told my father about this. I think when Hua looks back on this many years from now, he will agree with me. From that standpoint, he made a sacrifice. But he really wanted to improve mathematics in China, and he really thought that he would make more contributions. So, you have to ask me, it's all over now, but if he had stayed, by now, I suspect he would have done some very important work. Compare that result to what he actually did, and I think his decision at the time affected his academic development.

Who Is the Better Mathematician, Chow WeiLiang or Chern Shiing-Shen?

Ji Lizhen: I am also thinking of two more questions now. Are you familiar with Chow Weiliang?

Yang Chen-Ning: Chow Weiliang, yes, but not that much.

Ji Lizhen: I heard that André Weil once commented in a letter that he thought Chow Weiliang might be better at mathematics than Chern Shiing-Shen. Have you heard about this story?

Yang Chen-Ning: I didn't know he said that, but I know Chow Weiliang did foundational work in algebraic geometry.

Ji Lizhen: Yeah, Chow variety, Chow circle.

Yang Chen-Ning: I think because he had no students, his personality was maybe too … I think young mathematicians of Chinese descent should write some articles about this a bit more clearly.

Ji Lizhen: Yeah, wrote more about Chow Weiliang's contribution.

Yang Chen-Ning: I think both Chern and Hua thought that his contribution was very important, that it was a founding role in that field.

I know him, but not that familiar. Why? The main reason I know him is that I went to Stony Brook in 1966, and Simons became the dean after about a year or two, but before Simons became the dean, the president was desperately looking for a dean, and one of the candidates was Chow Weiliang. Chow Weiliang was asked to go to Stony Brook, so I also saw Chow Weiliang there. I don't know if he was not invited back by Stony Brook or if he was invited, but he refused to go there. I think we can find that out. You wrote a letter to the physics department and the mathematics department of Stony Brook. There must be records. But if you ask me, I think it's partly because of Chow Weiliang's personality and style.

Ji Lizhen: It's a little low-key, isn't it?

Yang Chen-Ning: He didn't seem to be interested in mathematics. He gave me that impression.

Ji Lizhen: Yeah, because he didn't seem to be as good at selling his work and sharing it with others.

Yang Chen-Ning: Well, he married a German girl after studying in Germany, and then he stopped doing math.

Ji Lizhen: Yeah, for a dozen years.

Yang Chen-Ning: Then he went back to China. And then, when Chern Shiing-Shen came back in 1946, he was probably encouraged to go back to mathematics, and that's when he did his work. And then he went to Johns Hopkins University, and Johns Hopkins thought a lot of him. But if you ask me, I don't think he was suitable to be a department chair, because he was kind of like … He just didn't have a passion for math; he didn't give you the impression that this guy had a passion for math, so he's not suitable for department chair, in my opinion.

Ji Lizhen: He might be like the ancient Chinese literati, living in seclusion in the mountains.

Yang Chen-Ning: There are very few people like him, but sometimes he did a very important job.

Influence of Yang Chen-Ning's Father on Hua Loo-Keng

Ji Lizhen: That's right. I was recently reminded of the upbringing of Hua Loo-Keng, who wrote an article about the unsolvable problem of quintic equations and then went to Tsinghua University. Later, he did Waring's problem, Waring's problem of number theory, and I think his work was heavily influenced by your father.

Yang Chen-Ning: His earliest articles were written at Northeastern University. They were all about Waring's problem, and it was my father who suggested that he do it. But he soon published in England, and his first articles were all on Waring's problem.

Ji Lizhen: Did your father ever talk about having some kind of relationship with him? Because your father was already a professor and Hua Loo-Keng was a student…

Yang Chen-Ning: Because when Hua Loo-Keng came to Tsinghua, I don't know who first noticed that he had written an article.

Ji Lizhen: It was Xiong Qinglai.

Yang Chen-Ning: But in the department, of course, Mr. Xiong was the head of the department, so I told Tang Peijing to find Hua because Tang Peijing seemed to know Hua. I had the impression that he did once, anyway. After he was hired, everyone thought that Hua should be an assistant because he could not be a graduate student. At that time, there were three of them. Mr. Xiong was in analysis, and Sun Guangyuan was in geometry. Do you know Sun Guangyuan?

Ji Lizhen: Yes, but I'm not familiar with him.

Yang Chen-Ning: He went to Chung-ang University. My father was a number theorist. He was an algebra guy. So naturally, Hua Loo-Keng came to my father, who gave him titles for his first few articles. You can read all these articles, and I have seen these articles because when I was at the University of Chicago, there was a very strong library, and I looked, and I saw these articles. But he soon moved on to other directions, and by the time he went to England, he was no longer working on the Waring's question.[66]

Ji Lizhen: When he first became famous, it was because he had won the prize for Additive Theory of Prime Numbers. I think that's related to Waring's question, and that's what made him famous.

Yang Chen-Ning: He went to Cambridge because … I don't know if it was Hadamard's suggestion or Norbert Wiener's suggestion.

Ji Lizhen: Yes, it was Norbert Wiener's suggestion. Let me ask you another question: What about Wu Wenjun?

Wu Wenjun and His Mathematical Work

Yang Chen-Ning: Wu Wenjun?

Ji Lizhen: Yeah, I saw a CCTV interview with him a few months ago, then I sent a link to the interview to some people, and some people didn't like it.

Yang Chen-Ning: Why is that?

---

[66]*AMS Notices* has a recent article about Dickson and his influence on his students, referring to three students, including Yang Chen-Ning's father and Hua Loo-Keng. Dumbaugh, Shell-Gellasch, "The 'Wide Influence' of Leonard Eugene Dickson," *Notices Amer. Math. Soc.* 64(2017), no. 7, 772–776.

Ji Lizhen: Because Wu Wenjun said something very interesting in it. He said some people were very good when they were young, and then they became ordinary later. He felt like he's not saying explicitly that people like him seem to be learning all the time, but we all know who the young man he's referring to is. It's René Thom. René Thom is a Fields Medal winner who pretty much went to school with him. Did you hang out with him much, with Wu Wenjun?

Yang Chen-Ning: As for Wu Wenjun, I have read some articles about him and what he said. I have the impression that, of course, I am not in this line of work. In fact, his contribution to the topology at that time was recognized, but he did not get the recognition that he should have; that is to say, he had a situation at that time, something he later said, that shocked the mathematical community in Paris, but I have never heard of anyone who engaged in topology in particular … It was my impression at that time that Serre, Borel, Thom and he were the most famous individuals in Paris at that time. But I think of the four of them; his name is now … All three have become very important: Thom, Serre and Borel. You didn't ask Borel about Wu Wenjun?

Ji Lizhen: No, I hadn't thought of it before. I didn't ask.

Yang Chen-Ning: I don't think it's discrimination because he's back, and he's Chinese. It's just natural that people don't pay as much attention to him because his language is different. I think it's quite natural. To put it another way, if he had stayed in France, I think it would have been better for his studies than it is now.

Ji Lizhen: And he could do very good things later on.

Yang Chen-Ning: Later, he did something very important in this area.

Lin Kailiang: Mechanization?

Yang Chen-Ning: Yes, mechanization.

Ji Lizhen: What do you think about it? Do you think it's important?

Yang Chen-Ning: Of course, it's important, not mechanization, but the relationship between the whole structure of mathematics and logic. It's very important work. I was interested in it for a while. It was a function that I wasn't sure how the computer world would value, but it was extraordinary. It was a very extraordinary mathematical problem.

Ji Lizhen: But his work is not quite the same as today's machine proofs. You know, the four-color problem, for example, was proved by computers. It's not quite the same as that.

Yang Chen-Ning: Yeah, let me see if that's true. Yeah, it's actually worth looking into … Now with today's understanding of computers, going back to look at the original Wu Wenjun work. I think the Chinese people should learn from the Japanese people to do this kind of historical work. I told the Japanese people this, and I said that Japanese people will fight for the history of physics, but no one in China will do this. Another thing that I think is also unfair is Wang Hao.

Ji Lizhen: Wang Hao, yes. He's famous in mathematical logic.

Wang Hao and Penrose Tiling

Yang Chen-Ning: I told you about this, didn't I? Penrose tiling was originally Wang tiling by Wang Hao.

Ji Lizhen: I don't know about that.

Yang Chen-Ning: I can find that out. Did I tell you that *Scientific American* has an article that makes it very clear? It was originally a logical problem, a mathematical logic problem, but Wang (Hao) created something called Wang tiling, which is a game. If the answer to the game is yes, there is a non-repeating puzzle that can be infinitely large, so if there is such a sentence, a theorem of logic has been proved. But the question is whether there could be such a game, if there is such a set of tiles that are non-repetitive and can be infinitely large. It was his idea, and it was a mathematical logic problem, but he didn't solve it. It was probably solved by one of his assistants, who said that it would take 365 different tilings to make an infinitely large one. Then the article came out and it went from over 300 to over 200 to two. And then Penrose came along, and he designed one, not the existing theorem, but he made one, and it became a game, so everybody knew about Penrose tiling, but nobody talked about Wang tiling.

Ji Lizhen: Well, actually, it meant that Penrose knew that that was Wang Hao's job.

Yang Chen-Ning: The professional people knew, but it's not known internationally. You put that article … I found this article later.

Ji Lizhen: I recently read an interview with Penrose. He said he had worked on it with his dad. Later, he was going to publish it in some magazine. He published it in a psychology magazine.

Yang Chen-Ning: Who do you mean?

Ji Lizhen: Penrose.

Yang Chen-Ning: Penrose?

Ji Lizhen: Penrose himself said that he and his father found this thing, and then he wanted to find a journal to publish it in, so he published it in a psychology journal. Yes, that's what he said. What do you think of Penrose?

Yang Chen-Ning: My opinion on Penrose?

Ji Lizhen: Yes. Do you know him well?

Yang Chen-Ning: There are two Penroses, two brothers. Oliver Penrose and Roger Penrose.

Ji Lizhen: Really? Are they both in math and physics? Are they doctors?

Yang Chen-Ning: I don't know which one is the elder brother and which one is the younger brother. One is a bit more mathematical and the other is a bit more physics-oriented.

Ji Lizhen: Really? There are two?

Yang Chen-Ning: I wrote an article that I think is not bad. It was influenced by an article written by Penrose and others. Penrose et al. wrote an article, and I wrote an article from their article. This article is one of my 13 most important articles. This Penrose is not the very famous Penrose.

Ji Lizhen: And the famous one is Roger Penrose?

Yang Chen-Ning: That's what people call Penrose. What Penrose did afterwards? I don't know.

Ji Lizhen: Penrose did the black hole; he was famous for the theory of relativity: singularity.

Yang Chen-Ning: Yes, you're talking about the famous Penrose.

Ji Lizhen: The famous one, and then I saw his video and sent the link to some people. I just said that Penrose is not bad and that he talks about that complex variable. But there are other people who disagree with me. They said that in general, listeners of Penrose's lecture understood only what they already knew, not what they didn't know beforehand. I had a question about Penrose, and another question was, do you know that there is an Institute for Advanced Study of the History of Science at Northwestern University? Have you ever been there?

Yang Chen-Ning: I don't know. Who was there?

Lin Kailiang: Qu Anjing.

Yang Chen-Ning: I know the place, but who is he?

Ji Lizhen: The old gentleman was famous.

Yang Chen-Ning: Did he study the history of physics or the history of mathematics?

Lin Kailiang: History of mathematics.

Yang Chen-Ning: What kind of work did he do?

Lin Kailiang: Mathematical astronomy. His teacher was Li Jimin.

Ji Lizhen: That's right.

Yang Chen-Ning: I don't know that person.

Ji Lizhen: I'd just like to ask you if you are willing to go to Xi'an and deliver a lecture at their university. He asked me to deliver this message.

Yang Chen-Ning: Willing to go to Xi'an. Is this person studying the history of science there?

Ji Lizhen: Yes. I'm going to visit the university, too.

Yang Chen-Ning: Who is he most interested in, and what does his research focus on?

Ji Lizhen: He used to study astrophysics. Now he studies Galois theory.

Yang Chen-Ning: Oh, Galois theory.

Ji Lizhen: Yeah, he studies Galois theory.

Yang Chen-Ning: So, he studies the history of mathematics.

Ji Lizhen: Yes, and he knows how much you stress the importance of the history of science. Because I'm going to Xi'an next week, I just thought I'd ask you if you are willing to go to Xi'an and when you could deliver a lecture at the university.

Yang Chen-Ning: Well, I think it's Galois theory. I admire Galois theory, but I haven't yet understood the best and most important thing about Galois theory, so I don't have a lot in common with him.

Ji Lizhen: Not much in common. Actually, I think they're …

Yang Chen-Ning: My father was very impressed by Galois. I never was. The most important thing about mathematical theories is that they all have one wonderful thing. But I haven't mastered the most amazing thing (about Galois theory) yet.

Ji Lizhen: How about inviting you to give a lecture without talking about that?

Yang Chen-Ning: No, I think Abel had proved that polynomials of degree five might not necessarily be solved in radicals before Galois.

Ji Lizhen: It can't be solved.

Yang Chen-Ning: But what was great about Galois was that you wrote down an equation, and he created a group structure from that equation, and with that group structure, he could tell you whether you could solve it or not, so it went a step further.

Ji Lizhen: This group theory is very important.

Yang Chen-Ning: So, of course, Abel had made a significant contribution, but the impact of his significant contribution, I don't think, can be compared to the impact by Galois because Galois made group theory important, and group theory covers everything. I believe that the next thing we need to do in high-energy physics, which is very important, is to have new symmetries. Where does this symmetry come from? String theory has been thinking about it, but it's been running around and hasn't hit on the right thing. I don't know what state it's going to be in 20 or 30 or 100 years from now. But I can tell you, I don't think it's impossible. It's human. This complexity of nature is beyond … I've always felt that we are limited. I think we're pretty good, but we're finite because we have a finite number of neurons.

Weng Fan: It's about time …

Ji Lizhen: Okay, thank you!

Yang Chen-Ning: Sorry, guys. Talk to you later.

The Eighth Interview

Time: June 20, 2019
Place: Yang Chen-Ning's home at Tsinghua University
Interviewer: Ji Lizhen
Recorder: Ji Lizhen
Collator: Peng Cheng, Wang Liping, Ji Lizhen

This interview was arranged after a conference held at Tsinghua University in memory of Zhang Shoucheng, so naturally, we talked about Zhang Shoucheng and his unexpected death. Yang Chen-Ning then talked about Chow Weiliang, his work, and his opportunity to be director of the mathematics department, etc. Mr. Chow Weiliang was considered one of the top mathematicians of the 20th century and had an influence on the work of mathematicians like Serre. The comparison between Hua Loo-Keng and Chern Shiing-Shen came up again. Then we talked about the Fields Medal and its influence, which mathematicians should be on it, and also how a science prize works and continues to be influential, like the Shaw Prize. As in every interview, Dr. Yang Chen-Ning made great comments and insights about scientists and their lives and work.

Zhang Shoucheng and His Deeds

Ji Lizhen: Last time, I listened to the last report of the meeting in memory of Zhang Shoucheng. I was very moved when I saw his notes. I didn't expect him to take notes so neatly. He took notes every day, including his own ideas.

Yang Chen-Ning: He was a very systematic person.

Ji Lizhen: Yeah, very systematic.

Yang Chen-Ning: Whose report did you listen to?

Ji Lizhen: It was from the last one, the one at 5 p.m.

Yang Chen-Ning: I didn't go. Was it Qi Xiaoliang?

Ji Lizhen: It was probably one of his students, now a colleague of his at Stanford, a former student of his.

Yang Chen-Ning: Qi Xiaoliang?

Ji Lizhen: Yes, it's Qi Xiaoliang.

Yang Chen-Ning: He was his best student.

Ji Lizhen: Yeah, and then he shared some of his manuscripts and stuff.

Yang Chen-Ning: One or two of his most important articles were co-written with Qi Xiaoliang.

Ji Lizhen: Really? Because I saw him start out very neatly. Later, I remember, he noted down something from 2018, which was not written as neatly as before, so I think maybe he was thinking differently. I was very surprised to hear that he had passed away because we had invited him to give some lectures in Sanya since we have a master lecture series in Sanya.

Yang Chen-Ning: What year was that?

Ji Lizhen: Two or three years ago, several years ago. He did a good job on his topological insulators (lecture), and Xue Qikun, vice president of Tsinghua University, joined him. After the lecture, I also sent a letter to him, inviting him to write a relatively popular article about his work, which was later published in our *Mathematics and Humanities*. Later, I was surprised to hear that this had happened.

Yang Chen-Ning: What year did you say he went to Sanya to give a speech?

Ji Lizhen: I can't remember, maybe two or three years ago; maybe three years ago.

Yang Chen-Ning: No, he passed away in December last year. Two years passed between that.

Ji Lizhen: Yeah, just for a while, because I didn't talk to him anymore. You see, when they spoke on Friday, several of them were also very moved and sad. Of course, I have heard a lot from the outside, and there were various reasons why he took this path.

Yang Chen-Ning: Well, here's the thing. He studied at Stony Brook. He was my Ph.D. student. He once studied in a program at Fudan or something, and he was still in his teens at that time; it was around 1980.

Ji Lizhen: That's right.

Yang Chen-Ning: As a result, Shanghai reached an agreement with Germany to send 100 young people from Shanghai to Germany. He took the exam and was selected. He got a master's degree in three years at the Free University in Germany, so he came to see me in Stony Brook. When he came to me, I gave him a research topic, and he came back a week later and said he didn't like it. What do you like to do,

Yang Chen-ning with Zhang Shoucheng (Right), taken in 2004

then? He said he likes doing this supergravity thing. Supergravity was coined by a colleague of mine named Peter van Nieuwenhuizen, and he said he wanted to work on this with Peter van Nieuwenhuizen. And well, I said in the United States no one says that you're not allowed to get another tutor, but I said that what he was doing was a little far from reality. You have to stand on your own two feet. So, I said that he should also learn a little condensed matter because that was more realistic. He accepted that, so he went with another professor, who is here this time, Steven Kivelson.

Ji Lizhen: Yeah, and he gave a talk, too.

Yang Chen-Ning: He is also at Stanford, but at that time he was in Stony Brook. He also attended Kivelson's class and co-authored an article. And he earned a Ph.D. Also, his doctoral dissertation was written with van Nieuwenhuizen, and then he came up with a job offer, but I told him not to work in Chicago. If he went to Chicago, his work would be purely theoretical; if he went to Santa Barbara, where there's a Nobel Prize winner, I said he could do condensed matter physics. So he went there.

Ji Lizhen: Oh, so that's how it all happened.

Yang Chen-Ning: He went there. I wonder if you know a man named Wen Xiaogang?

Ji Lizhen: I've heard of him.

Yang Chen-Ning: Wen Xiaogang is a professor at the Massachusetts Institute of Technology and he's doing very well. He and Zhang Shoucheng were there at about the same time, both working on a very abstract approach to string theory and condensed matter physics. They co-authored an article at Santa Barbara.

Ji Lizhen: Really?

Yang Chen-Ning: But then they went different ways.

Ji Lizhen: Oh, that's what it looks like.

Yang Chen-Ning: Zhang Shoucheng gave up string theory completely, while Wen Xiaogang continued to work mainly on string theory because he was Witten's student, and he got his Ph.D. with Witten, so he is still there. Zhang Shoucheng gave up his work with van Nieuwenhuizen.

Ji Lizhen: Oh, that's right.

Yang Chen-Ning: So, Zhang's most successful work was done between 2004 and 2006, right when the work on topological insulator was published. Since it was written, there have been three main people working on this: he, an American named Charles L. Kane, who was also a theorist, and a Russian who was an experimenter. The Russian and Charles L. Kane were all here this time, so it was universally acknowledged that it should be credited to all three of them.

Ji Lizhen: It deserves a Nobel Prize.

Yang Chen-Ning: Yes, he has been nominated many times. It's not clear why he hasn't won. I think now that he's gone, those two will definitely get it.

Ji Lizhen: I think so.

Yang Chen-Ning: I know an article about that. I can email it to you.

Ji Lizhen: Okay, please email it to me.

Yang Chen-Ning: In 2013, he started another fund. Because he was smart and he knew a lot of people at Stanford who invested in funds, he invested very successfully. So, after that, he went back to do physics, and he went back to invest in funds.

Ji Lizhen: Did the foundation fail or something? Finally, it was …

Yang Chen-Ning: His invested in a lot of bitcoin funds, so because of that, it probably lost a lot of money. But that would not be what led to his suicide, because it was not all his money, so it was not a problem. My guess is that he was a frequent visitor to China.

Ji Lizhen: The FBI came after him?

Yang Chen-Ning: The FBI is investigating him. They don't know what's going on.

Ji Lizhen: What are they investigating? No idea?

Yang Chen-Ning: It's a pity.

Ji Lizhen: I think it's easier for him when he's gone, but you can imagine how hard it was for his wife and kids? I think maybe when he decided to go, he had a hard time figuring it out. You know, how much it will affect his family when he leaves the world. I think a smart guy like him should have taken that into consideration.

Yang Chen-Ning: This time his daughter has come back. He has a son and a daughter. The son is in the United States, but he didn't come this time. His wife came with her daughter.

Ji Lizhen: I was at East China Normal University on Saturday, and the president of East China Normal University Press showed me his wife's article on Tomb-Sweeping Day in memory of him.

Yang Chen-Ning: Because his wife is from East China Normal University.

Ji Lizhen: Yes, she graduated from East China Normal University in mathematics. That's why I'm surprised now.

Yang Chen-Ning: Who do you know in his mathematics department?

Ji Lizhen: The dean is Tan Shengli.

Yang Chen-Ning: You know him, right?

Ji Lizhen: Yes, I know about Tan Shengli.

Yang Chen-Ning: He was in Germany. Wasn't he in Germany?

Ji Lizhen: Yes, he visited Germany. I'm not sure which university.

Yang Chen-Ning: He got his Ph.D. in Germany?

Ji Lizhen: I'm not sure. I went to East China Normal University this time and had a chat with the president of their press. One of them has already sorted out half of the things from our interview last time. After sorting it out, Li Ping and I will look at it carefully and then show it to you. We plan to publish it this year or early next year.

Yang Chen-Ning: Is it a book or something?

Ji Lizhen: A book.

Yang Chen-Ning: Is there adequate stuff for a book?

Ji Lizhen: We'll include the first interview that's been published. You see, the last conversation we had was 50 pages, and the next conversation was 50 or 60 pages. If you multiply 40 or 50 pages five or six times, there are hundreds of pages.

Chow Weiliang, His Family and His Work

Yang Chen-Ning: I'll tell you one more thing about this. Maybe I haven't told you. Recently, I've heard that Chow Weiliang is doing very well, and Mr. Chern (Shiing-Shen) has also told me so many times. When I went to Stony Brook in 1966, the president of Stony Brook at that time was John Toll, a fellow of mine who knew how to do things, and he became the president. It was when he became president that he pulled me out of the Institute for Advanced Study and brought me to Stony Brook. When I went there in 1966, I found that there was no dean in the mathematics department, so he was actively looking for a dean, and one of the deans he found was Chow Weiliang.

Ji Lizhen: Chow Weiliang was a candidate.

Yang Chen-Ning: So, he came down from Johns Hopkins to Stony Brook to give some speeches, and I had dinner with him.

Ji Lizhen: Really?

Yang Chen-Ning: Talk to him. I knew him. Do you know him?

Ji Lizhen: I haven't met him face to face. I've only heard of his theorem.

Yang Chen-Ning: He's one of those who doesn't talk much. He's a gentleman.

Ji Lizhen: Very low-key.

Yang Chen-Ning: His wife is German, and for some reason, he doesn't want to come to Stony Brook.

Ji Lizhen: Really?

Yang Chen-Ning: I suspect he doesn't want an administrative job.

Ji Lizhen: That's possible.

Yang Chen-Ning: You can see that it's not easy for him to do administrative work. Do you know that Chow's family is a big family in Tianjin?

Ji Lizhen: Yes, I heard that. His family has a lot of property.

Yang Chen-Ning: I'm not sure. Was his grandfather or his father a big entrepreneur?

Ji Lizhen: Probably, because after he got his degree in Germany, he came back home to run his family's business, and he left after a decade or so.

Yang Chen-Ning: Mr. Chern asked him to come back and do mathematics.

Ji Lizhen: Because it was the War of Resistance against Japanese Aggression, and then his family's company might have had some problems, so he came back to do mathematics, which is great, I know.

Yang Chen-Ning: Do you understand why his theorem is important?

Ji Lizhen: Well, there are two aspects. The first time I heard about his theorem was in an algebraic book, and I heard about Chow's theorem. I said, I've never heard of this guy. He's like an ancient Chinese. What he was saying was this: if you look at the complex projective space, like this complex space, how do you determine that there is a subspace in it? There are two ways: you are given an equation, and the equation can define a subspace of it, and the functions that define the equation can be holomorphic functions, or polynomials. You see, there are many more holomorphic functions than polynomials, right? And we know that some functions are not polynomial, transcendental, transcendental functions. Then he had a very big theorem called Chow's theorem; that is, any closed subvariety in a complex projective space must be algebraic. A subset defined by a holomorphic function must be defined by a polynomial.

Yang Chen-Ning: When was this theorem introduced?

Ji Lizhen: It was around the 1950s.

Yang Chen-Ning: The 1940s, right?

Ji Lizhen: Maybe the 1940s. I can't remember. I'll have to look it up. Yeah, around the 1940s. Then Mr. Serre, who came to see you last time at Tsinghua, came up with a more general theory, but the main essence was in the Chow theorem.

Mathematical Work of Serre and Chow Weiliang

Yang Chen-Ning: Serre in France?

Ji Lizhen: Jean-Pierre Serre.

Yang Chen-Ning: How do you spell it?

Ji Lizhen: S-E-R-R-E, Jean-Pierre Serre. He is generally regarded as one of the best mathematicians alive.

Yang Chen-Ning: Serre was probably a few years younger than me.

Ji Lizhen: Yes, younger than you.

Yang Chen-Ning: I saw him at the Institute for Advanced Study. He played table tennis with people all day.

Ji Lizhen: Yeah, he always thought table tennis was good.

Yang Chen-Ning: You're saying that Serre's stuff has something to do with Chow Weiliang?

Ji Lizhen: Yeah, Serre was working on functions of several complex variables or something in the 1950s, and he had a very famous article called GAGA (Algebraic Geometry and Analytic Geometry).

Yang Chen-Ning: But I thought Serre mainly studied topology.

Ji Lizhen: Yeah, Serre was famous for topology in his twenties, and then in the 1950s, he moved to functions of several complex variables, algebraic geometry, and then number theory. He was also involved in the solution of Fermat's theorem. And Serre did a lot of work, because Serre had a very famous paper called GAGA, which was French for analytic geometry and algebraic geometry, and he popularized Chow's theorem, which is one of Chow's great works. The other thing is that in topology there are cohomology and homology theories, and the Chow ring (Chow Weiliang ring) is also very important now in algebra and arithmetic, and I think both will be handed down.

Yang Chen-Ning: So, he's not just known for that work, but for other works, too?

Ji Lizhen: Yeah, Chow variety, too, because Chow variety is actually very important. How do we describe a space that is rich? There are a lot of subspaces in it, which are more interesting spaces. Why is our plane geometry interesting? (It's because in plane geometry.) There are straight lines, triangles and lots of interesting spaces. So, if we study algebraic geometry, algebraic varieties, and subspaces defined by polynomials like this, it's a very important question as to what the structure of all the spaces that they form is, and that's another great piece of work by Chow.

Yang Chen-Ning: So, nobody knows why he didn't do later?

Ji Lizhen: He did it later.

Yang Chen-Ning: Did he?

Ji Lizhen: He did it all the time.

Yang Chen-Ning: You say he was still doing it at Johns Hopkins University?

Ji Lizhen: Yeah, he's been doing it all along. He just didn't do it as much as he had before. He is still doing it.

Yang Chen-Ning: Does he have any students?

Ji Lizhen: I don't think so. I haven't heard so. We did an article about him, and we tried to get in touch with his family, and then we got in touch with his wife and daughter, an old professor at Johns Hopkins University, and we got some pictures.

Hua Loo-Keng and Chern Shiing-Shen: Different Styles and Mathematical Achievements

Yang Chen-Ning: Well, as mathematicians, Chow is differnt from other mathematicians in personality and research methods. I think when you compare Hua with Chern, if they were to compete in the Olympics,

Hua would be better than Chern. Hua had done a lot of things, and of course, he was very smart, and he could solve a lot of problems, but he has not solved a problem of great impact. Mr. Chern, of course, had an opportunity to obtain his intrinsic proof, because first of all, he heard and appreciated intrinsic proof, and he had been working on it himself. Everyone wanted to generalize to higher dimensions, but he didn't know what the integrand was. It turned out that when André Weil found the integrand, he told him and gave him this article. It was a very difficult article to read, and I don't think anybody reads it now, but it was about taking a higher-dimensional manifold, cutting it up into little pieces and then putting it back together, and it was a very hard thing to do. Chern said, "I don't need this," so he found a new path, which is of course very important.

Ji Lizhen: I heard that Mr. Hua lost several opportunities when he went to England. He didn't learn anything from Hardy.

Yang Chen-Ning: Not from whom?

Ji Lizhen: Hardy.

Yang Chen-Ning: I thought he was there to work with Hardy.

Ji Lizhen: No.

Yang Chen-Ning: No cooperation?

Ji Lizhen: Yeah, the time was relatively short. If only Hua had really learned something from a master like Hardy, his work would be a little deeper. And then he went to Princeton, and he thought it was a big pity that he didn't learn some real kung fu from masters like Weyl and Siegel.

Yang Chen-Ning: My impression was that he was competing with Siegel at the time.

Ji Lizhen: Not really. It is said so in China, and also in the biography of Hua Loo-Keng written by Wang Yuan, which is actually wrong because Siegel was almost one of the greatest mathematicians in the 20th century, and people like André Weil admired him very much. André Weil has a passage in his complete works where he said that mathematicians of their generation, like him, and like Weyl … It was a very meaningful thing to do to comment on Siegel's work. And then they wrote a bunch of articles, and I looked them up, and some of them were just to get a deeper understanding of Siegel's work, and Siegel was really, really amazing.

Yang Chen-Ning: When you talk about this, I don't know if I've told you, but a lot of mathematicians are very excited about the 100 mathematicians of the 20th century …

Ji Lizhen: Yes, there are rankings.

Yang Chen-Ning: If you are more or less interested in physics, you can't do it.

Ji Lizhen: Why not?

Difference Between Mathematics and Physics and the List of Top Physicists

Yang Chen-Ning: This is to say that physics and mathematics are basically different.

Ji Lizhen: Why not physics?

Yang Chen-Ning: In terms of value, physics and mathematics have different values, and they have different problems to solve. I think the two fields are very closely related, but they are very different. Let alone a graduate student, can you name 100 most important physicists of the 20th century?

Ji Lizhen: No, I can't.

Yang Chen-Ning: No, I don't think most of the professors can, either.

Ji Lizhen: So, you think people can name the top ten, right?

Yang Chen-Ning: Yes, that's likely. I think university professors, of good universities in the United States can name 50 people, as for 100 …

Ji Lizhen: The average person's list is too small.

Yang Chen-Ning: It depends on how valuable people's contributions are. This means that when you make something, it has value, but in physics, it has no value after it has been surpassed by others. After being surpassed, mathematics is like this: you can be surpassed by others, but the thing you do is still valuable, because it has something to do with art. Gauss had a theorem. How many ways did he have to prove it?

Ji Lizhen: Yes, the quadratic reciprocity law, eight ways.

Yang Chen-Ning: Each of his proofs, you said you have proved, but are other proofs worthless? Actually this is not true. It has value because it's a work of art. Physics is not about art. If you discover what has been done by previous researchers in physics, your work will be forgotten by others as well as by yourself.

Ji Lizhen: Yeah, that's true.

Yang Chen-Ning: If you prove a theorem in mathematics, it's like drawing a picture that you can hang on the wall and always find interesting. Physics is not like that.

The Nobel Prize as the Most Successful Prize

Ji Lizhen: Yes, it is. I have a question now. You see, there are a lot of new prizes, such as the New Breakthrough in Science award in the United States, which comes with a prize of 3 million dollars. There is a Future Science Prize in Chinese mainland, and the Shaw Prize in Hong Kong SAR. There are also many local prizes. In mathematics, there are Fields Prizes, but among all those prizes, I still think the Nobel Prize is the best. The question is, why is the Nobel Prize so successful? What do you think is the reason that the Nobel Prize is so important, influential and incomparable?

Yang Chen-Ning: I think about, for example, why is the mathematics prize not as influential as the Nobel Prize? There are several reasons for this. One reason is that the Nobel Prize happened to start at the beginning of the 20th century. Physics and chemistry made great discoveries at the beginning of the 20th century, and that makes the Nobel Prize what it is. Do you know what the first Nobel Prize in Physics was given to?

Ji Lizhen: Was it an Australian?

Yang Chen-Ning: X-rays.

Ji Lizhen: Oh, X-rays, I forget their name.

Yang Chen-Ning: His name is Röntgen.

Ji Lizhen: Oh, Röntgen, right.

Yang Chen-Ning: Because everybody knows about these when they go to the doctor. Everybody understands X-rays. Then you take Jacques Salomon Hadamard, for example, who proved the theorem of prime numbers.

Ji Lizhen: People don't understand and don't care.

Yang Chen-Ning: It's different when you talk to people and they don't understand. That's one thing. It happened to be in the early decades of the 20th century that both mathematics and physics were advancing by leaps and bounds. That's why the Nobel Prize was born. So, biology came a little bit later, and it didn't become important until the 1950s with the papers by James Dewey Watson and Francis Harry Compton Crick. But all in all, because there are three Nobel Prizes that belong to science and two that do not, it's the Nobel Prizes for science that make the Nobel Prizes for non-science so famous. And the other thing is, it's fairly equal; it's basically fair.

Ji Lizhen: It's pretty much fair.

Yang Chen-Ning: There are mistakes. It's not as fair as mathematics. Mathematics is a little bit fairer.

Ji Lizhen: Really? Is mathematics fairer?

Yang Chen-Ning: Basically, it is.

Ji Lizhen: It's fair.

Yang Chen-Ning: And it has a long history. You know, in fact, in the early 1900s, there was a big mathematical prize. I think Hilbert and Poincaré both won it. I think it was called …

Ji Lizhen: The Bolyai Prize, isn't it?

Yang Chen-Ning: No, it's a German name. The prize (money was in Mark), and during the second time …

Ji Lizhen: Oh yeah, it went down.

Yang Chen-Ning: After the first time, the Mark currency was not worth much, so the prize was lost later.[67]

---

[67]The name of the prize is the Wolfskehl Prize. It's related to Fermat's Last Theorem. Poincaré won the founding prize, the Bolyai Prize, in 1905, and Hilbert won the second award.

Ji Lizhen: Yeah, I've heard about that. The proceeds from that prize are used to hire people.

Yang Chen-Ning: The Nobel Prize has a lot of money. It also has problems. At the beginning of the Nobel Prize, it had a lot of purchasing power, but then they invested too conservatively. In the 1950s, that's when I got it …

Ji Lizhen: The amount of money has been relatively small?

Yang Chen-Ning: The purchasing power is at its lowest.

Ji Lizhen: Oh, really?

Yang Chen-Ning: Then they changed their investment policy to invest in real estate, so it went up again after the World War II.

Ji Lizhen: Oh, that's right.

Yang Chen-Ning: But now it's a problem again.

Ji Lizhen: Really?

Yang Chen-Ning: The problem is that they have too many activities now. They have meetings, they have organizations, and they have a museum. It was all too expensive. So, whether or not the Nobel Prize will be well managed in the future, I'm not sure yet.

Ji Lizhen: Yes, I heard that the Nobel Foundation seems to be very unhappy about the Nobel Prize in Economics.

Yang Chen-Ning: No, it's not.

Ji Lizhen: It doesn't count?

Yang Chen-Ning: The Nobel Prize in Economics is not awarded by the

Nobel Foundation; it is given by the Bank of Sweden, but people think it is a Nobel Prize.

Influence of Prestigious Awards

Ji Lizhen: Yeah, everybody puts it all together like that, and the Nobel Foundation feels like it's losing out. Some scientists have suggested whether it is possible to have a Nobel Prize in mathematics, but the Nobel Foundation says absolutely not, because they feel that the Nobel Prize has already established a high reputation and they don't want to do the same thing again, so I think it's not easy to ever have a mathematics prize. On the other hand, I told you about Borel in my email last time, and he said that this kind of prize was both good and bad because those who got it were happy, but those who didn't get it were depressed. The other day, Borel came to give an academic report, and three other colleagues and I had dinner with him. Borel talked about this matter, and he said that it might be better to abolish many awards because there are many harms associated with them.

The next day, one of my colleagues met me in the hallway. He smiled and said that Borel might have been talking about himself the night before because Borel had done a lot of work, but Borel hadn't won any big prizes. I also introduced an article to you last time. A person analyzed the influence of the Nobel Prize after winning it. That is to say, there are different influences. There is also an article about mathematics. It seems that after winning the Nobel Prize, some people's achievements and creativity have decreased a lot. So, what do you think? Is it a good award? Does the good outweigh the bad?

Yang Chen-Ning: Well, if you ask me, I think that even if the two sides cancel out, it's still positive. It's what gets people interested in doing scientific research and helps encourage people to do it. As for the idea that there are some inequities and even some errors, there are, but on average, I think it is positive.

Ji Lizhen: It's good, yeah.

Yang Chen-Ning: But now there is a new phenomenon. It is because of the income generated by this money that there are suddenly countless billionaires. China has some billionaires now, and they all hold different kinds of prizes. Then, after these awards, of course, they can't (compare with the Nobel Prize), because the Nobel Prize was established earlier, so it occupies a (historical) place, so you can't squeeze it out, but now you have all kinds of these … I think all these prizes are competing, foreign countries are competing, and China is competing. Especially China now sees it very clearly, but there's also a part of China now that I think is getting to understand. It's not a prize like the Nobel Prize; it's going to be, let's say, the latest prize. I forget what it's called.

Ji Lizhen: It's called Frontier.

Yang Chen-Ning: That's the Future Prize.

Ji Lizhen: Oh, it's called the Future Prize.

Yang Chen-Ning: There was another one recently.

Ji Lizhen: Another one?

Yang Chen-Ning: The most recent one is called—I forget what it's called, and its purpose is to encourage young people, so it's mainly aimed at young people. But there's a lot of them, not one or two, but tens of hundreds of them, so this one serves a different purpose.

Ji Lizhen: You said there were more prizes?

Yang Chen-Ning: That's right.

Ji Lizhen: It's kind of like a scholarship, so this is good.

Establishment of the Qiu Shi Award

Yang Chen-Ning: Yes, actually, I've been involved in this award for many years. It's called the Qiu Shi Award.

Ji Lizhen: Oh, the Qiu Shi Award.

Yang Chen-Ning: The Qiu Shi Award has been given for more than 20 years, 25 years.

Ji Lizhen: Who is it usually given to? I've only heard of the Qiu Shi Award.

Yang Chen-Ning: The Qiu Shi Award started with a rich man in Hong Kong SAR, a billionaire. He is no longer alive. His name is Cha Chi-Ming.

Ji Lizhen: Does it have anything to do with Jin Yong?

Yang Chen-Ning: They are both surnamed Cha, and they are distant relatives, but they know each other, though not well.

Ji Lizhen: That's the way it is.

Yang Chen-Ning: This family had a great painter during the Qing Dynasty.

Ji Lizhen: Really?

Yang Chen-Ning: This old gentleman, Mr. Cha, he's about ten years older than me. He's a billionaire. Around 1992, he felt that he wanted to help Chinese mainland develop science and technology. He started making money in textiles. His textile factory was in Nigeria, and it was the biggest textile factory there. Then, he did real estate, and then he did finance. He has a very good relationship with Chinese mainland. Why is that? He has money because his father-in-law has money. His father-in-law's surname is Liu, and he's from Changzhou, Jiangsu province.

Ji Lizhen: The textile industry in Changzhou is very good.

Yang Chen-Ning: In the early years of the era of the Republic of China, there were some rich men in Jiangsu. His family was one of them, Mr. Liu. So, Cha Chi-Ming was Liu's son-in-law. During the war, this (Mr.) Cha helped his father-in-law develop his textile business in Chongqing. At the time of liberation, Mr. Liu stayed on the Chinese mainland, and his son-in-law and daughter went to Hong Kong SAR, where they made a fortune. Mr. Liu stayed in Nanjing, where he knew very well the famous general of the People's Liberation Army.

Ji Lizhen: I don't know about the general in Nanjing, either. Is it the commander?

Yang Chen-Ning: Probably the commander of the Nanjing Military Region.

Ji Lizhen: Yes, I seem to remember the name, but I can't remember. Is it Xu Shiyou?

Yang Chen-Ning: Xu Shiyou. Mr. Liu is very familiar with Xu Shiyou, so of course, this (Mr.) Cha has a good relationship with Chinese mainland. In 1992 and 1993, he wanted to donate money to help Chinese mainland with its scientific and technological experiments, so he turned to Mr. Chern (Shiing-Shen) because he had a son who was still at Stanford, had a finance company, and Mr. Cha had a house in Palo Alto, so he knew Berkeley, Palo Alto and Stanford as well. He donated a lot of money, and then he met Mr. Chern, and through Mr. Chern, he found me, and then at Berkeley, Lee Yuan-Tseh. So, he found Mr. Chern in mathematics, found me in physics, found Lee Yuan-Tseh in chemistry, and found a man called Kan Yuet Wa in biology. Kan Yuet Wa is a very important biologist and is still alive and a few years younger than me. He is a professor at UC San Francisco, and he almost won the Nobel Prize.

Ji Lizhen: Really? All four are good.

Yang Chen-Ning: So, the four of us began to be his advisers, and the first award was in 1994 in Beijing.

Ji Lizhen: Who was the mathematics award given to?

Yang Chen-Ning: The first award was given to ten people, four of whom were involved in the atomic bomb project, four in the satellite project, and two others, one of which was for the replantation of severed limbs.

Ji Lizhen: Oh, limb replantation, transplant.

Yang Chen-Ning: Limb retransplant was invented by the Chinese.

Ji Lizhen: Really? I didn't know.

Yang Chen-Ning: Well, if your foot is broken, you have to connect it with a lot of blood vessels and nerves. Foreigners are not careful enough. He developed this. I think this is Nobel Prize-level work, but the Nobel Prize doesn't award this kind of work, as it is more clinical.

Ji Lizhen: There's no theory to it.

Yang Chen-Ning: The Nobel Prize is for the study of principles, so we gave it to this team. Well, in the beginning, it was given to people who had made great contributions to China's national defense.

Ji Lizhen: In the field of national defense.

Yang Chen-Ning: Of course, this cannot be continued.

Ji Lizhen: No, there are not so many people.

Yang Chen-Ning: After 20 years, it has become this "Qiu Shi Fund" (Qiu Shi Science and Technology Foundation). Now, Mr. Cha is no longer alive, but his son is doing it. It's called "Qiu Shi Fund." Every year it gives out prizes.

Ji Lizhen: So, you're still there, as a consultant?

Yang Chen-Ning: I present the trophies to the award winners every October. I will retire after this year's award ceremony in October. Now, some new consultants will replace us.

Ji Lizhen: I have a question. In all the years that you have been giving awards, what do you think has been satisfactory and successful about these awards, and what have been some of the negative aspects? The ones that need to be improved.

Yang Chen-Ning: Unlike the Nobel Prize, the "Qiu Shi Award" is specifically designed to promote the development of science and technology in China.

Ji Lizhen: Oh, it's for China.

Establishment of the Shaw Prize

Yang Chen-Ning: And my other thing is the Shaw Prize.

Ji Lizhen: Yes, the Shaw Prize.

Yang Chen-Ning: The Shaw Prize had nothing to do with this old man. Sir Run Run Shaw came to me. Why do I know him? Because he's from the Chinese University (of Hong Kong).

Ji Lizhen: Alumni?

Yang Chen-Ning: Shaw College Trustee, so I've known him since I became a professor at the Chinese University (of Hong Kong). He asked me, and he was not even 90 years old at the time, but he asked me why the Nobel Prize was so successful, and I wrote him a letter, and then I heard nothing from him, so I thought he was probably …

Ji Lizhen: Probably not so serious.

Yang Chen-Ning: The most important thing I said in my letter is the truth of success. One is that the awards are fair, one is that the money is more, and the other is that it lasts a long time. Sir Run Run Shaw didn't answer, so I didn't think he was interested. But a dozen years later, when he was 90, he and his wife came to me and said he was interested now, and that he was going to set this all up, so the Shaw Prize was established. The purpose of the Shaw Prize is to work with …

Ji Lizhen: To compete with the Nobel Prize?

Yang Chen-Ning: To compete with the Nobel Prize, if you ask me, quite successfully. One way to be quite successful is to give …

Ji Lizhen: One or two mathematics prizes?

Yang Chen-Ning: One for mathematics, one for astronomy and one for biology. Why are mathematics and astronomy included? Because the Nobel Prize doesn't have either.

Ji Lizhen: No Nobel Prize for those. No.

Yang Chen-Ning: Why did you give one to biology? There is a Nobel Prize in biology. It's because after we discussed it, we thought that biology has a lot of varieties now, and it's developing very fast, so we can give it an award, too. So, it's been 20 years since we first gave out these three prizes.

Ji Lizhen: It's very successful now.

Yang Chen-Ning: Very successful. Why is it successful? Because there is already a dozen …

Ji Lizhen: Run Run Shaw Prize and then the Nobel Prize, right?

Yang Chen-Ning: I won the Shaw Prize first, then the Nobel Prize a few years later, so …

Ji Lizhen: Your judgment was successful.

Yang Chen-Ning: It was a success. Then, of course, you know mathematics. It's all about … It's just that, I think, we all think, the Shaw Prize hasn't been given to the wrong person so far.

Ji Lizhen: The last time David Bryant Mumford came to talk to me about it, we met at a conference and it was a very funny thing he said. He said he had taken the Shaw Prize with Wu Wenjun and he explained why he had received the Shaw Prize and he thought maybe the selection committee wanted to use him to push Wu Wenjun. What do you think? He told me so. Mumford said so himself.

Yang Chen-Ning: He got it, didn't he?

Ji Lizhen: Yes.

Yang Chen-Ning: When did he get it?

Ji Lizhen: I can't remember exactly. He won the Shaw Prize in the same year as Wu Wenjun.

Yang Chen-Ning: The main reason is that we, the jury, were right. In the beginning, it was Atiyah for a long time, so if Atiyah were the chairman of the committee, he certainly wouldn't choose a bad person. So that's the most important thing; that's what the five people of the judging panel need to really understand.

Ji Lizhen: To understand mathematics. There's basically someone I know of, Langlands, because Langlands hadn't won a big prize before.

Yang Chen-Ning: Langlands? We gave him this award.

Ji Lizhen: Yeah, I think it's great.

Yang Chen-Ning: It turned out that he gave a speech, a semi-colloquial speech, and you had no idea what he was talking about. I think he's probably the worst speaker. What was the British man's name? He's a number theorist. He won the Shaw Prize. His wife is Chinese.

Ji Lizhen: Richard Taylor.

Yang Chen-Ning: He gave a great speech. It was very easy to follow.

Ji Lizhen: He proved Fermat's theorem with André Weil. And Kontsevich won the same prize. He was Russian. He was in Germany. Most of the people who won the Shaw Prize were very successful.

Yang Chen-Ning: The Shaw Prize is different from other prizes because it has three. Like the Abel Award, there's only one, and so is the Crafoord Prize.

Ji Lizhen: One or two.

Yang Chen-Ning: I remember they took turns. It was this year, next year and the year after. The Shaw Prize is more like the Nobel Prize.

Ji Lizhen: Yes, it happens every year.

Yang Chen-Ning: Now Sir Shaw was gone, and his wife took over, and suddenly his wife was gone. So now the Run Run Shaw Group's money is managed by a man named Chern, and that man's name is Chern Weiwen. I have retired. Sir Shaw's wife was still the chairman when I retired. Two or three years later, at this time the year before last, his wife was no longer there. After his wife died, we were all a little worried about what would happen to his money. Then this guy, Chern, came along. He wasn't from the academic world, but he was very interested in keeping it going, so it looked like he could keep it going.

Ji Lizhen: Because the Nobel Prize has a foundation and its money is always there, but the Shaw Prize does not have that guarantee?

Yang Chen-Ning: Yes.

Ji Lizhen: That's a little bit dangerous because if the people in charge of the money change their minds, then there's no more prize.

Yang Chen-Ning: Yes, there is a danger, and there is no way around it. But because it's so much money, it's much more than this …

Ji Lizhen: More money than a Nobel Prize?

Yang Chen-Ning: Sir Run Run Shaw left a very small amount of money to support the prize, so there will be no immediate crisis. But you're right. In fact, when Sir Shaw was still here, we on this committee told him … No, it was when his wife was here. We told her we were going to turn it into a foundation, like the Nobel Prize, but she never would. And the reason she didn't, I think, was because she said that if she was going to start a foundation, she would have to find someone to run it, and she couldn't find that person right away, so the foundation wouldn't survive. But in this state, she was gone, so we didn't know what was going to happen. As a result, she gave the money to this person surnamed Chern in her will. I barely know this person named Chern.

Ji Lizhen: He doesn't have any descendants? Sir Run Run Shaw had no children?

Yang Chen-Ning: Yes, Run Run Shaw had children. He was successful. He was one of two brothers, and he was the younger brother, and both of them made their fortunes in Malaysia and Singapore. After becoming rich, they separated around the time of liberation, with his brother staying in Malaysia and Singapore, and Run Run Shaw coming to Hong Kong SAR to develop himself and exchange their sons.

Ji Lizhen: Sons exchanged? Why the swap?

Yang Chen-Ning: This is what China used to do … It was that Run

Run Shaw took his nephew to his house and sent his own children to his brother. As a result, he lived on the largest inheritance, and there was nothing for his family.

Ji Lizhen: Really?

Yang Chen-Ning: There was a memorial service when Run Run Shaw was away, and I went there, and they introduced me. His son held my hand, and his son was not close to his father.

Ji Lizhen: Yeah, because he didn't raise him. Why change it? As for this, I don't know about the tradition of changing sons in China. Why change them?

Yang Chen-Ning: It's a very complicated tradition in China. Some of them are called "hugging and giving."

Ji Lizhen: It usually means that if the other person has no children, I give mine to him, but why should I change it if the other person has one?

Yang Chen-Ning: This is one kind. There is another kind. For example, for me, I was given to my uncle.

Ji Lizhen: Really? Your uncle doesn't have any children, does he?

Yang Chen-Ning: No, it's not that he doesn't have children; it's that he feels like…

Ji Lizhen: So, the child can grow up healthy?

Yang Chen-Ning: That's right.

Ji Lizhen: That's superstition.

Yang Chen-Ning: God blesses them more in this way.

Ji Lizhen: Really? This way?

Yang Chen-Ning: There exists this very strange idea.

Ji Lizhen: God have more mercy on them for that.

Yang Chen-Ning: But I think it's good for him and his brother, his adopted son, because…

Ji Lizhen: It's harder to manage your own kids, isn't it?

Yang Chen-Ning: Yeah, because your relationship with your nephew can be easier in some ways than your relationship with your son …

Ji Lizhen: Communication?

Yang Chen-Ning: Less emotional influence.

Ji Lizhen: Business is better. Oh, that's the reason. I think the Shaw Prize has gone through such a long time that, even if there is no guarantee, someone else will do the same in the future.

Yang Chen-Ning: The Shaw Prize is not a problem now. I think the main reason is that the Shaw Prize is only a small part of a large amount of money. Because the Nobel Prize has become independent, it is good, but if it is not well managed …

Ji Lizhen: The money won't come from anywhere.

Yang Chen-Ning: That's right.

Ji Lizhen: Do you think the judges of the Shaw Prize are similar to the Nobel Prize? Basically, who wins the prize has something to do with who is important on the judging panel.

Yang Chen-Ning: Our method is copied from the Nobel Prize, but there are some differences. In the Nobel Prize's judging panel, there are five people in a discipline, and the five people must be …

Ji Lizhen: Members of the Academy, right?

Yang Chen-Ning: You have to be a Scandinavian; that's Dane, Norwegian, Swede, and I don't know if it includes the Finns. They should come from among the three countries, so we don't have that.

Ji Lizhen: You're from all over the world?

Yang Chen-Ning: Our judges are from all over the world, and we invite very important people.

Ji Lizhen: The last time when the Fields Medal was awarded, there was a problem. People were complaining about Jacques-Louis Lions, who was the president of the IMU, the International Mathematics Union.

Yang Chen-Ning: What was he called?

Ji Lizhen: L-I-O-N-S, Jacques-Louis Lions.

Yang Chen-Ning: Does he have a son called Pierre Lions?

Ji Lizhen: Yes, the father was president of the International Mathematical Union, and then he chose a committee for his son to sit on.

Yang Chen-Ning: I've heard that rumor.

Ji Lizhen: There are a lot of people who have strong opinions. They say that if you choose the committee, it will be good for your son. Has that ever happened to the Shaw Prize?

Yang Chen-Ning: No, I think we've been careful up to this point, and the judging panel of the Shaw Prize is appointed. That's quite right.

Ji Lizhen: Yes, I think that's important for an award.

Yang Chen-Ning: Of course, there is a problem here. Some mathematicians may think that algebraic geometry is the most important, more important than topology. Others may not think so. But this is inevitable.

Ji Lizhen: It's unavoidable.

Yang Chen-Ning: But not all of these five people are the same.

Ji Lizhen: Yeah. When I wrote to you last time, I read that the winners of this year's prize are related to probability, and the recent Fields Medal winners seem to be related to probability, too, so I got the idea that if we can look at the whole development of mathematics from the situation of the winners …

Yang Chen-Ning: I think the fact that these winners deal with probability shows quite a number of important mathematicians perceive this field to be increasingly important.

Ji Lizhen: Actually, I think it's an interesting thing. I told you about that article last time, which is 100 years of physics and the Nobel Prize in Physics. This seems to mean that they write articles that sometimes don't describe major advances in physics. Later, I looked it up. It seems that there was no mention in the article. It just discussed the work of the laureates and didn't mention anything else.

Nobel Prize Winners and Ph.D.

Yang Chen-Ning: Of course, there is another important consideration for the physics prize, which is to award the prize to researchers with theoretical or experimental achievements. I think there is a bit of a deviation here. For example, a few years ago, the prize was awarded to a researcher in applied sciences. At that time, I thought that the discovery of this application could be quickly turned into a big industrial application. Later, it turned out to be wrong. If it were wrong, many people would think that you did not give the award properly. There are some such awards. But there is another difference between physics and mathematics. There are some results in physics that are not understood until many years later. The most important one is called Josephson junction. Brian David Josephson, I think it was in the early 1960s, was a graduate student, and he was English. He wrote an article saying that there was a special phenomenon in superconducting things, and he guessed it, and it turned out to be right.

Ji Lizhen: Really?

Yang Chen-Ning: And it had a very important application, so he won the Nobel Prize. I think it was 1961 or 1962. When I was at the Institute for Advanced Study, Princeton University, there was a Nobel Prize winner in theoretical physics, 20 years older than me. You must have heard of him; he was named Wigner.

Ji Lizhen: Wigner, yes. He wrote a very famous article called "Unreasonable Effectiveness of Mathematics," right? Oh, he did representation theory. I was wrong.

Yang Chen-Ning: The earliest paper on finite-dimensional representation.

Ji Lizhen: He did it, right.

Yang Chen-Ning: He wrote it. He's Dirac's brother. Dirac told Wigner that he thought he could work on it. Dirac didn't do a lot of mathematics himself, but Wigner was very good, so I think this mathematical prize recognizes him first. So, Wigner and I, and a guy about my age in the Physics Department of Princeton, the three of us brought in this guy Josephson, who was just getting his Ph.D.

Ji Lizhen: That's quite something.

Yang Chen-Ning: We had been talking to him all morning, trying to figure out what was going on with him. When he left, the three of us agreed that if he were a Princeton grad student, we wouldn't approve his Ph.D.

Ji Lizhen: Why not?

Yang Chen-Ning: Because he couldn't articulate his stuff.

Ji Lizhen: Really?

Yang Chen-Ning: He guessed, but after he guessed, he couldn't explain the theory clearly. He hasn't explained it clearly yet.

Ji Lizhen: People don't understand it yet?

Yang Chen-Ning: It's just that physics is a little different from mathematics in this regard because, in the end, physics is an experiment. There are many strange things in this experiment. I have been doing it for many years and I have some understanding, but it is not the most basic understanding.

Ji Lizhen: That's right.

Renormalization and the Nobel Prize

Yang Chen-Ning: I'll give you the most important example, which is called renormalization.

Ji Lizhen: I've heard the name renormalization.

Yang Chen-Ning: What is the renormalization problem? It is that you usually read mechanics and you have to solve a lot of exercises. For example, if an exercise has three degrees of freedom, if you write a Hamiltonian operator which has Q, P, Q_i, P_i, say $i = 1, 2, \ldots, 5$, that's five degrees of freedom. But if you have an electromagnetic field, it has a particularly unlimited number of degrees of freedom because when you transform it, each component of the frequency is an independent variable, so the electromagnetic field has an infinite number of degrees of freedom. It is still not clear about infinite degrees of freedom in mathematics.

Ji Lizhen: Really?

Yang Chen-Ning: Mathematicians don't know, and physicists certainly don't know. You know, of course, if there are three degrees of freedom, P_1, P_2, P_3, $P_1^2 + P_2^2 + P_3^2$, then add one function about $Q(Q_1, Q_2, Q_3)$. Isn't that a dynamic probability? An electromagnetic field is not three dimensions; however, it is an unlimited number of dimensions, especially the unlimited multiple degrees of freedom, and usually it cannot be explained. But if it has a parameter in its latent part, and when that parameter is small, you can expand in powers of that parameter, which we'll call α, and its value is $1/137$. You know about that; it's called a fine structure constant.

Ji Lizhen: Is this Planck's constant?

Yang Chen-Ning: No, it's called the fine structure constant. It's $e^2/(4\pi\varepsilon_0 ch)$, e, h, c. You should know all about it. This $e^2/(4\pi\varepsilon_0 ch)$ is a dimensionless quantity, and its value is equal to $1/137$. Why such an important thing has this value is still unknown. But with this value, because it's small, if you take this particular problem to the freedom limit, that potential v has this fine structure in it, and you will expand it to a different exponent of this variation.

Ji Lizhen: Oh, and you get the first coefficient?

Yang Chen-Ning: Yes, so this came out in the 1930s, this whole electromagnetic field formula, and the most important person at that time was Dirac, so he was able to calculate it, and this fine structure constant was calculated to the lowest order, and it was consistent with the experiment.

Ji Lizhen: This is called renormalization?

Yang Chen-Ning: It's not renormalization yet. This is the first step. So, let's come to step two; step two is dispersive.

Ji Lizhen: Yes, dispersive ones require renormalization.

Yang Chen-Ning: But it wasn't able to be solved in the 1930s. At that time at Heisenberg, a lot of people were working on it, and I thought it was weird because you had the lowest order, and it was perfect, but if you added one more order, it got infinite, so it was called infinite difficulty. When I was a graduate student, the most important thing was the infinite, and it became infinite because it had unlimited degrees of freedom, and every one of them contributed, adding up to the infinity. After the war, the first and most important contribution was called renormalization, and later these three men won the Nobel Prize. It is said that the thing you calculated is infinity, but that infinity is not infinite. If you modify the fine structure constant, it is not infinite. This is called renormalization. That is, you can make it when you sort through those infinite things and that's what was done in 1946, 1947.

Ji Lizhen: Is that what the Feynman path integral is? Is he one?

Yang Chen-Ning: Yeah, Feynman is one, Schwinger is one, and Tomonaga Shinichirō is one.

Ji Lizhen: Oh, that's when Dyson didn't win the prize …

Yang Chen-Ning: Actually, I don't think that's true at all, because what Dyson did was very important because not only did he prove Feynman, Feynman guessed, but all three of them could only do level 2, not level 3, and not level 4. Dyson was able to do all the orders, and he told you how to do it, so he was this …

Ji Lizhen: Well, I just said there's a list of the top 100 mathematicians. Dyson has to be on it. He's a very important one.

Yang Chen-Ning: Of course, he is very good. I know him very well. I admire his ability to solve mathematics.

Ji Lizhen: Really?

Yang Chen-Ning: I think few people can compete with him.

Ji Lizhen: Really? He is very good at mathematics.

Yang Chen-Ning: But here's the neat thing. They three won the Nobel Prize. In the 1930s, the first term was calculated, and because they three could calculate the second term, they won the Nobel Prize.

Ji Lizhen: Yeah. No one got any prize afterwards for doing the calculations.

Yang Chen-Ning: Dyson came and told you that you could calculate not only the second, but also the third, the fourth, and now the fifth term.

Ji Lizhen: Really?

Yang Chen-Ning: The accuracy is one billionth.

Ji Lizhen: Really? That's pretty impressive.

Yang Chen-Ning: It's amazing, because someone who knows this stuff, a graduate student, you can lock him in a room and not allow him to use anything, just calculate, and within a week or two, if he's good, he can come up with a result that's (only) one in a billion (off) the experimental result.

Ji Lizhen: That's pretty accurate.

Yang Chen-Ning: But no one understands why. And now it looks like this: if you continue to calculate, it is still the same because you have no reason, one in a billion is already right, the experiment is very accurate, and the calculation is also very accurate. This is …

Ji Lizhen: That's amazing.

Yang Chen-Ning: This means that there is no such problem in mathematics. This is nature, so I think this may be what humans will never understand.

Ji Lizhen: That's right. You just said that no one won the Nobel Prize after calculating the first and second terms.

Yang Chen-Ning: That's right.

Ji Lizhen: That's a pity.

Yang Chen-Ning: Then there were no more (Nobel Prizes) in this field.

Ji Lizhen: What was the reason? Is it that the Nobel Prize jury didn't think it was important? But is it actually important to the development of physics itself, or not so important?

Yang Chen-Ning: The reason is that there are so many problems, and because in this case, people think that if you get the second one right, the third one might not be so important. And then there is another problem. If we continue to calculate it, no one knows whether it is a convergent series or a non-convergent series. So, Dyson wrote a paper saying that it's not convergent, and that it's best to just calculate a few terms, and then don't do it anymore.

Ji Lizhen: It's not worth it. It diverges.

Yang Chen-Ning: I actually added a little bit to this one.

Ji Lizhen: Really? Do you still make this stuff?

Yang Chen-Ning: The way I pointed it out to Dyson is that Dyson tells you how to do it, not just the second term, the third term, the fourth term, but he outlines a method. I went and looked into it, and I found that there was a loophole in the method, and the loophole was that by the time you got to the seventh term, Dyson's method was not clear.

Ji Lizhen: That's the way it is.

Yang Chen-Ning: Well, very few people noticed it, but Dyson noticed it.

Ji Lizhen: It shows that people still know not much about nature.

Yang Chen-Ning: There are very few people who really understand how Dyson's renormalization works. Most people, especially those who study physics, are not as persistent as those who study mathematics.

Nobel Prize Winners and Religion

Ji Lizhen: The man you just mentioned, the man who did super-conductivity in the 1950s, said he was not very clear about the theory at the beginning. How did his work develop later?

Yang Chen-Ning: Who is it?

Ji Lizhen: You said there was a young man.

Yang Chen-Ning: Josephson?

Ji Lizhen: Yes. How did his work develop?

Yang Chen-Ning: He stopped doing it.

Ji Lizhen: He stopped doing it?

Yang Chen-Ning: He turned to religion.

Ji Lizhen: Really?

Yang Chen-Ning: He's still here.

Ji Lizhen: So, in a sense, this is one of the more failed examples of the Nobel Prize?

Yang Chen-Ning: Well, I don't think it's a failure.

Ji Lizhen: Really? Didn't you say his work is important?

Yang Chen-Ning: Because his work is right, it is compatible with experiments, and he can make practical things with this method, so he has made contributions. As for his lack of clarity, that's another thing.

Ji Lizhen: Why did he go into religion? That's a bit strange.

Yang Chen-Ning: He's not the only one.

Ji Lizhen: Really? Are you saying that some of the other Nobel Prize winners have gone into religion?

Yang Chen-Ning: Do you have such people in mathematics?

Ji Lizhen: I heard there was a Russian who did a kind of arithmetic geometry and went mad. Because the way it works in mathematics is you look at all the integers, and you look at the prime numbers, 2, 3, 5, 7, and then he adds infinity. He says his idea is to put everything together, and infinity and this kind of finite together is a very important arithmetic solution, and then it turned out he went crazy.

Yang Chen-Ning: Is it just wishful thinking?

Ji Lizhen: Yeah, and then he said he saw God in infinity.

Yang Chen-Ning: Where did he come from?

Ji Lizhen: Russia.

Yang Chen-Ning: A Russian?

Ji Lizhen: Yes.

Yang Chen-Ning: He was doing something important, right?

Ji Lizhen: It's important.

Yang Chen-Ning: Very important?

Ji Lizhen: Yeah, that's what Zhang Shouwu does.

Yang Chen-Ning: What does it have to do with Zhang Shouwu?

Ji Lizhen: The work that Zhang Shouwu started to become famous for was in this aspect. It was based on that man's work.

Yang Chen-Ning: Following that person's work?

Ji Lizhen: That guy created a theory called …

Yang Chen-Ning: It's number theory?

Ji Lizhen: It's called Arakelov geometry. The man's name is Suren Yurievich Arakelov. Normal number theory only deals with one prime number, two, three, five, seven, and then Arakelov talks about connecting the complex geometry together. One of the important things Zhang Shouwu did was to connect arithmetic and complex geometry, which was very important.

Young Chinese Mathematicians

Yang Chen-Ning: Are there some students of Zhang Shouwu who are doing well in the Langlands program now?

Ji Lizhen: Yes, he has a student named Zhang Wei.

Yang Chen-Ning: Zhang Wei seems to have won an award recently.

Ji Lizhen: One of Zhang Wei's classmates is Yun Zhiwei. He is also very famous. Yun Zhiwei is not Zhang Shouwu's student, but MacPherson's student and also works with the Fields medalist Wu Baozhu from Vietnam. They are doing very well.

Yang Chen-Ning: Is this Wu from Vietnam?

Ji Lizhen: Yeah, Wu Baozhu is Vietnamese.

Yang Chen-Ning: But with his name, he might be of Chinese descent?

Ji Lizhen: That's right. Last time, Yang Le also told me about it. He said, "Look, we Chinese have to work harder. Look, the Vietnamese have won the Fields Medal. The Iranians have won the Fields Medal. The Brazilians have won the Fields Medal. But we Chinese haven't."

Yang Chen-Ning: Why?

Ji Lizhen: I don't know. But here's the thing. There are a lot of people who think that many of the current winners aren't doing very well. So, that's the question I wanted to ask. The Nobel Prize has always had a good reputation and good quality, but a lot of mathematicians think that the Fields Medal today is not good enough. Last time Borel said to me (he knows me better. I'm sort of one of his more recent students) that the current Fields Medal is not a good idea. He said that before 1986, we all knew who had won the prize, and after that, now no one knows.

Yang Chen-Ning: Do you think a lot of people feel that way?

Ji Lizhen: Yeah, a lot of people think so.

Yang Chen-Ning: Then maybe they should cut down numbers of winners and do not award four each time.

Ji Lizhen: Yeah, it's political. They're fighting over it. Your people are going up; my people are going up. For example, many people were very critical of the winner selection in the previous award. For some people, you could not tell what important work they had done, but it only said that they had done extensive and deep work. Such general statements are hard for people to believe. If you take the other winners, and you say very clearly why they won the prize, William Paul Thurston and Alain Connes, each of them had a very important job, and that's what Borel was saying before, we all knew which one of us deserved the prize. That's what I'm saying about the Fields Medal, and it's hard to say what will happen to the award if it keeps going like this. So, I think there are a lot of people who are uncomfortable. That's what Borel said: the people who get the prize are happy, and the people who don't get the prize are not so happy.

Yang Chen-Ning: Who decides the overall policy of the Fields Medal? It has a committee, doesn't it?

Ji Lizhen: Yes, the president of the International Mathematical Union appoints a selection committee to judge the prizes.

Yang Chen-Ning: So, is this called the IMU (International Mathematical Union)?

Ji Lizhen: Yeah, IMU.

Yang Chen-Ning: Does the chairman of the IMU have any influence?

Ji Lizhen: The president? Yeah, he's very influential.

Yang Chen-Ning: The president of the IMU is elected?

Ji Lizhen: Yes, it should be.

Yang Chen-Ning: Who chooses?

Ji Lizhen: It's a mathematical union. Every time before the mathematicians' conference, they have to meet to choose the next one. It's like this.

Yang Chen-Ning: Yes, the next one will be selected by the United States, Germany and Europe.

Ji Lizhen: Yes, they are all represented. We have representatives at the IMU.

Yang Chen-Ning: Who are the Chinese?

Ji Lizhen: I think China is Tian Gang now.

Yang Chen-Ning: Tian Gang?

Ji Lizhen: Yeah, he should be from the IMU.

Yang Chen-Ning: Because he was chosen by the state?

The Fields Medal Winners at the 2002 Beijing Congress of Mathematicians

Ji Lizhen: Yes, they represent the country. They have to be citizens of that country. Let me tell you something. In 2002, the World Congress of Mathematicians was held in Beijing. The Fields Medal was awarded to two people: one was Russian, the other was French. The Russian got drunk after he won the prize, and he went to the Institute for Advanced Study at Princeton, and he died two years ago—no, a little over a year ago.

Yang Chen-Ning: What was his name?

Ji Lizhen: Vladimir Voevodsky.

Yang Chen-Ning: What kind of stuff did he make?

Ji Lizhen: What he did was related to Chow Weiliang's work. It's a Chow variety. Yeah, the work he was doing was actually directly related to Chow Weiliang's Chow ring. I heard that after he won the prize, he found out that one of his theorems might be wrong, and then he spent a lot of time fixing it. After that, he lost confidence in the whole work.

Yang Chen-Ning: Let me ask you a question. Nowadays, mathematicians often have to write one or two hundred pages of a theorem, so they usually write a short article before they write it. There are many loopholes in the short article, so there may be a quarrel.

Ji Lizhen: Yeah.

Yang Chen-Ning: And it's getting worse and worse, isn't it?

Ji Lizhen: Yes.

Yang Chen-Ning: It wasn't like that a century ago.

Ji Lizhen: Yes. William Paul Thurston gave him the Fields Medal at the time without fully proving it, and he never wrote it again.

Yang Chen-Ning: Who?

Ji Lizhen: Thurston.

Yang Chen-Ning: Is Thurston still like that?

Ji Lizhen: Thurston didn't write it later, someone else did, so if they gave the prize later, as I said about Voevodsky, they were careful to wait until it was completely written.

Yang Chen-Ning: Is Thurston's work related to Poincaré's?

Ji Lizhen: Yes, the topology of three dimensions.

Yang Chen-Ning: Extending Poincaré conjecture a little bit?

Ji Lizhen: Yes, three-dimensional classification. Someone told me that he didn't actually write it when they gave him the prize, and he stopped writing it after he got the prize. So, when they gave the prize to Voevodsky, they delayed it for a few years, hoping that he would write it first, but then he found out that he had written it wrong, so there was still a big problem.

Yang Chen-Ning: But did others make up for it?

Ji Lizhen: No, he made up for it himself.

Yang Chen-Ning: Made up?

Ji Lizhen: Yes, he did. He spent a lot of time making amends. But after making amends, he lost faith in mathematics. Because it hit him so hard, he drank too much. The other guy who won the Fields Medal in Beijing basically stopped doing mathematics, so that was a big problem.

Could I ask you again? I mentioned in my email that I am going to write a book called *The Swordsman in Mathematics*. What do you think of it? Because there is a very famous novel called *The Swordsman*. My idea is to tell the story of some mathematicians and their work, and to reflect on the development of mathematics and the interaction of mathematicians.

Yang Chen-Ning: I think the general idea is good, but whether you succeed or not depends on you.

Ji Lizhen: The question is what to write. It's not good to have gossip; it's not good to have too much, and it's also good to have mathematical content.

Yang Chen-Ning: I think especially comparing the similarities and differences between mathematics and physics is a good topic, so I think you should listen to this.

Ji Lizhen: Yeah, I think it's useful and interesting. Actually, I want to write a book about it. Last time I discussed it with you, I wrote an article comparing Hilbert and Poincaré. What do you think?

Yang Chen-Ning: I have read it. It's very interesting.

Ji Lizhen: Was it okay? So, here's my plan. I'll have it translated into Chinese. What do you say?

Yang Chen-Ning: Good. Will you translate it yourself?

Ji Lizhen: I can't translate it; I can't write in Chinese. I need someone to translate it directly into Chinese.

Yang Chen-Ning: Lin Kailiang knows people who can translate.

Ji Lizhen: Yeah, he does. Do you find it valuable to publish in English?

Yang Chen-Ning: Where do you plan to publish it?

Ji Lizhen: One possibility is in *ICCM Notices*.

Yang Chen-Ning: *Mathematics, Science, History and Culture*, isn't it?

Ji Lizhen: Not *Mathematics, Science, History and Culture*.

Yang Chen-Ning: You mean he has an English-language magazine?

Ji Lizhen: Yes, *ICCM Notices* has an English version.

Yang Chen-Ning: That's not a mathematical study?

Ji Lizhen: No, it's more like a review article.

Yang Chen-Ning: Is it like the *Mathematical Intelligencer*?

Ji Lizhen: Yeah, kind of like that.

Yang Chen-Ning: It's in English?

Ji Lizhen: Yes, it's in English.

Yang Chen-Ning: Why isn't Wang Liping here this time? The girl from Higher Education Press.

Ji Lizhen: She went to work as a lawyer. She left.

Yang Chen-Ning: Where is she now?

Ji Lizhen: Nanjing.

Yang Chen-Ning: What is she doing in Nanjing?

Ji Lizhen: In a law firm in Nanjing.

Yang Chen-Ning: Is that what she does?

Ji Lizhen: Yeah, she's a lawyer.

Yang Chen-Ning: How can she become a lawyer? She didn't study law.

Ji Lizhen: She studied mathematics, but after six months of self-study, she passed the law exam with high marks. She passed the law exam and became a lawyer. So, it's a shame that she was the head of the Higher Education Press's academic works bureau, and now she's gone. But she continues to work on your books, and even though she's a lawyer, she thinks she still wants to work on some of them. She said she wanted to say hello to you and that she couldn't come this time because she was in Nanjing. It was a surprise to me, too, because she did a really good job in academic publishing, but she left publishing.

People's lives change a lot, and there are a lot of things that you don't expect. But she would certainly do well as a lawyer.

Yang Chen-Ning: Okay.

Ji Lizhen: Thank you!

People Notes

A

Niels Henrik Abel (1802–1829) was a Norwegian mathematician. He proved the impossibility of solving quintic equations with root expressions, and made important contributions to power series, elliptic function theory, Abel integrals, and a class of algebraic equations (Galois groups with commutations). His work in the field of elliptic function theory and Abelian integrals directly contributed to the work of Bernhard Riemann and Karl Weierstrass and greatly influenced the development of complex analysis and algebraic geometry. The Abel Prize was established by the Norwegian government in 2002 to commemorate the 200th anniversary of Abel's birth and has been awarded annually since 2003 by the Norwegian Academy of Science and Letters, to honor a mathematician of depth who has made outstanding contributions. Another purpose of establishing the prize is to make up for the lack of a Nobel Prize in mathematics. Abel's motto, "Learn from the masters, not from their students," is particularly instructive on how to cut through the clutter in an age of information overload.

Andreas E. Z. Alföldi (1895–1981), a noted expert on ancient history, was on the faculty of the Institute for Advanced Study in Princeton from 1955 to 1981. He was an expert on the history of ancient Rome (especially the frontiers of the Roman Empire), the rituals and insignia associated with imperial royalty, and the religious culture of the Roman Empire. He received numerous honors in his lifetime, including the Gold Cross of King George II of Greece (1936), the Medal of the Hungarian Archaeological Society (1943), the Huntington Medal of the American Numismatic Society (1965), the Cross of Honour of the Federal Republic of Germany (1972), and the Medal of Honour of the Republic of Austria in Science and Art (1975).

Vladimir I. Arnold (1937–2010) was a Russian mathematician and dynamicist. His research interests include the theory of dynamical systems, classical and celestial mechanics, differential equations, singularity theory, real and complex algebraic geometry, symplectic geometry, tangential geometry, fluid mechanics, variational method, differential geometry, potential theory, mathematical physics, function superposition theory, combinatorics, and history of mathematics. One of his most important achievements was with Andrey Nikolaevich Kolmogorov and Jürgen K. Moser, establishing the KAM theory of dynamical systems. Arnold, together with his mentor Kolmogorov, received the Soviet Union's Lenin Prize in 1965. He received the inaugural Crafoord Prize awarded by the Royal Swedish Academy of Sciences in 1982 and the Wolf Prize in Mathematics in 2001. Arnold is also known for his outspoken comments. For example, he has candidly evaluated Hilbert's 23 questions and the recent Fields Award winners' levels of scholarship.

Michael Francis Atiyah (1929–2019) is an English mathematician. His research fields include algebraic geometry, topology, algebra, analytics, and mathematical physics. In 1963, he worked with the American mathematician Isadore M. Singer, and proved the Atiyah-Singer index theorem. Atiyah was awarded the Fields Prize in 1966 and, together with Singer, the Abel Prize in 2004. He was elected a Fellow of the Royal Society in 1962 and became its president in 1990. He was awarded a knighthood in 1983. He was one of the few great mathematicians in the world who wrote well and gave very inspiring speeches. Atiyah was more than willing to give his own views on the various branches of mathematics; his articles on general topics in the field of mathematics are ideal for students to read.

B

Alan Baker (1939–2018) was an English mathematician who mainly studied number theory and made major breakthroughs in the branches of Diophantine approximation and transcendental number theory. By estimating the linear lower bound of the logarithm of algebraic numbers, he gave the upper bound of all groups of Thue equation, Mordell's equations, and hyperelliptic equations. Independently of H. M. Stark (Harold Mead Stark), he proved Gauss's conjecture for imaginary quadratic fields of the class number 1. That is, there were only nine of them known to Gauss, and there was no tenth field. Baker's book *Transcendental Number Theory* (1975) became a contemporary classic. He won the Fields Medal in 1970 and was elected a Fellow of the Royal Society in 1973.

Alexander Beilinson (1957–) is a Russian mathematician and professor at the University of Chicago. His research interest covers representation theory, algebraic geometry, and mathematical physics. In 1981, Beilinson announced that, together with Joseph Bernstein, he had proved the Kazhdan-Lusztig conjectures and Jantzen conjectures. He was awarded the Ostrovsky Prize of Switzerland in 1999. In 2017, he was elected a member of the National Academy of Sciences. He was awarded the Wolf Prize in Mathematics in 2018, and he received the Shaw Award in 2020. He was one of the best students in the prestigious Gelfand Discussion class in Moscow. He and Drinfeld now run a Gelfand seminar at the University of Chicago.

Eric Temple Bell (1883–1960) was an American mathematician born in Scotland. He made significant contributions to analytic number theory. He received his PhD from Columbia University in 1912 and was president of the American Mathematical Society from 1931 to 1933. Bell published about 250 scholarly articles. For his work *The Analysis of Arithmetic* (1921), he received the American Mathematical Society's Bocher Memorial Award in 1924. His books *Algebraic Arithmetic* (1927) and *The Development of Mathematics* (1940) became benchmarks in the field, the latter outlining in clear and concise language what Bell considered to be the most important trends in mathematics. Bell also wrote a number of popular mathematical science books, such as *The Master of Mathematics* (1937), *Mathematics: The Queen and Servant of Science* (1951), and *Fermat's Last Theorem* (1961). These books were often criticized by historians of mathematics for a variety of reasons, valid and otherwise. He also wrote and published quite a few novels under the pseudonym John Tyne.

Ludwig Bieberbach (1886–1982) was a German mathematician who received his Ph.D. from the University of Gottingen and taught at the Universities of Konigsberg, Basel, Frankfurt, and Berlin. He was a member of the Prussian Academy of Sciences. His main research interest covered function theory, differential equations and children's science. In 1916, he put forward the conjecture of univalent function coefficient, which was not confirmed until 1984 by the American mathematician Louis de Branges. He rose to fame in 1911 when he solved Hilbert's eighteenth problem on crystal groups in his professorship paper. He founded the infamous Nazi mathematical journal *Deutsche Mathematik* in 1936. After World War II, he was fired due to these historical issues and devoted himself to writing books, some of which are excellent and still appreciated and used by many mathematicians.

Salomon Bochner (1899–1982) was an American mathematician born in Krakow (present-day Poland) in the Austro-Hungarian Empire. He taught at the University of Munich in Germany and at Princeton University and Rice University in the United States. He was also a member of the National Academy of Sciences. In 1932, Bochner wrote his *Lectures on Fourier Integrals*, which contained Bochner's theorem on positive definite functions and the generalized Fourier transform as a precursor to the theory of generalized functions. Bochner created works in differential geometry, the theory of functions of multiple complex variables, and probability theory, and other fields. The Bochner formula is named after him and has had a great influence on both contemporary differential and complex geometry. He also wrote extensively on philosophy and the history of science.

Niels Henrik David Bohr (1885–1962) was a Danish physicist who was one of the first to apply quantum theory to the theory of the structure of atoms. He was a leading figure in the development of quantum mechanics and the founder of the Copenhagen School. He was elected a member of the Royal Danish Academy of Arts and Sciences in 1917. In 1918, Bohr published "On the Quantum Theory of Line-Spectra", elaborating on the idea of correspondence principle, which at that time became a bridge from classical theory to quantum theory. Bohr was awarded the Nobel Prize in Physics in 1922. In 1927, he proposed the principle of parallelism to explain quantum mechanics, which is one of the important achievements of the Copenhagen school.

Ludwig Eduard Boltzmann (1844–1906) was an Austrian physicist and one of the main founders of statistical mechanics. He received his doctorate from the University of Vienna in 1866. Boltzmann was one of the first scientists in Europe to understand the importance of Maxwell's theory of electromagnetism. In 1868, he generalized Maxwell's theory of the distribution of energy and derived the law of equal distribution of energy, known as Maxwell-Boltzmann's law. In 1872, he gave a mathematical expression for the thermodynamic entropy and the configuration distribution of gas molecules, known as Boltzmann's equation. In 1877, he further clarified the statistical properties of the second law of thermodynamics and was the first to use probability to explain the fixed law. At the turn of the 19th and 20th centuries, his statistical mechanics successfully explained a series of important discoveries in atomic physics and fluctuations in Brownian motion. He was a genius, but somewhat mentally troubled and, therefore, was likened to the famous painter, Van Gogh.

Enrico Bombieri (1940–) is an Italian mathematician. His research interests include number theory, algebraic geometry, partial differential equations, function theory of many complex variables, finite group theory, and differential geometries. His main achievements in number theory are to improve the large sieve method, to get the median common formula of the error term of the distribution of prime numbers in arithmetic series, and later to invent important methods to improve the exponential sum and Riemann ζ function estimation. He was awarded the Fields Medal in 1974. Bombieri was also a fine painter and worked on the design of the Mathematics School building at the Institute for Advanced Study in Princeton.

Armand Borel (1923–2003) was a Swiss-American mathematician. His research interests included algebraic topology, algebraic groups and arithmetic subgroups, Lie groups and Lie algebras, and automorphic forms. His most important contribution was to lay the foundation for the theory of linear algebraic groups. Borel was a member of the National Academy of Sciences and the French Academy of Sciences. He received the Brouwer Prize of the Royal Mathematical Society of the Netherlands in 1978 and the Lifetime Achievement Award of the Steele Prize of the American Mathematical Society in 1991. He was a very serious and responsible man who did everything with the precision of Swiss watch.

Émile Borel (1871–1956) was a French mathematician. He was elected a member of the French Academy of Sciences in 1921. In 1928, he helped found the Poincare Institute and served as its director until his death. Borel combined Georg F. L. P. Cantor's point set theory with his own knowledge to establish a theory of functions of real variables. His most famous work was the formulation of the finite coverage theorem and the generalization of measures from finite intervals to a larger class of point sets (i.e. Borel measurable sets), which established the basis of measure theory. His work on the *Theory of Divergent Numbers* won the prize of the French Academy of Sciences. In 1909, he introduced the concept of countable sets of events, which filled the gap between classical finite probability and geometric probability, and he also proved a special case of the strong law of numbers.

Satyendra Nath Bose (1894–1974) was an Indian physicist. He was president of the Indian National Academy of Sciences in 1950, and was elected a fellow of the Royal Society in 1958. Bose's most important contribution was the establishment of a kind of quantum statistics, known as Bose statistics.

Raoul Bott (1923–2005) was a Hungarian-American mathematician. He contributed to the fields of algebraic topology, Lie group theory, characteristic class theory, foliate structure theory, K theory, and index theory. In particular, he applied Harold Marston Morse's theory to the study of homotopy groups of Lie groups, which led to the Bott periodicity theorem, and thus led to the new field of topological K theory and even non-transformational geometry. Bott was the recipient of the American Mathematical Society's Veblen Prize (1964), the Steele Prize for Lifetime Achievement (1990), and the Wolf Prize in Mathematics in 2000. Two of his students won the Fields Medal, Stephen Smale and Daniel Gray Quillen, and some were very prominent mathematicians, such as Robert Mac Pherson. Macpherson, along with

Michael F. Atiyah, Friedrich Ernst Peter Hirzebruch, and Isadore M. Singer, formed the famous Gang of Four, a name given by Yau, a favorite among them.

Nicolas Bourbaki was the collective pen name chosen in the mid-1930s by eight (or nine) young mathematicians in France, whose founders included French Claude Chevalley, André Weil, Henri Cartan and Jean Dieudonné, among others. After World War II, they were joined by Samuel Eilenberg, a Polish-American. The group's original purpose was to write a rigorous textbook on mathematical analysis, but it grew to include an introduction to many branches of algebra and mathematical analysis from an axiomatic point of view, as well as topology. Bourbaki's work began in 1939 with *Éléments de mathématique* Volume one. *Éléments de mathématique* (now more than 30 volumes) became the classic reference work on the foundations of modern mathematics.

Luitzen Egbertus Jan Brouwer (1881–1966) was a Dutch mathematician. He emphasized mathematical intuition, rejected Georg F. L. P. Cantor's discussion of the real infinite, denied the absoluteness of the law of excluded middle, and established a constructivist system of mathematics, including a constructible continuum. Brouwer was regarded as the founder and representative of intuitionism. He made profound and unique contributions to topology, especially algebraic topology, such as Brouwer's fixed point theorem. Without his writings and a group of colleagues around him, topology might not have developed as well as it has.

C

Elie Cartan (1869–1951) was a French mathematician. Cartan was a major influence on the development of mathematics in the 20th century. His mathematical work could be broadly divided into three main categories: the theory of Lie groups and Lie algebras, the theory of differential equations, and differential geometry. The generalized space he proposed was the predecessor of the concept of fibrogen and the unification of Klein geometry and Riemannian geometry. He was Chern Shiing-Shen's true mentor.

Augustin-Louis Cauchy (1789–1857) was a French mathematician and member of the French Academy of Sciences. He was a professor at École Polytechnique, at the University of Paris, and at the University of Turin. His mathematical achievements covered many fields, and he established the foundation of complex analysis and real analysis.

Chen Kuo-Tsai (1923–1987) graduated from Southwest Associated University in 1946, went to the United States in 1947, and received a Ph.D. from Columbia University in 1950. Chen Kuo-Tsai's early work was mainly on group theory and the theory of links. He was best known for his work on iterative integrals, power series, and cohomology of cyclic spaces.

John Douglas Cockcroft (1897–1967) was a British experimental physicist. In 1931, together with Ernest Thomas Sinton Walton, he first converted lithium transmutation into helium, ushering in a new era of nuclear reactions using artificially accelerated particles. For this work, they shared the 1951 Nobel Prize in Physics. During World War II, Cockcroft worked on radar, anti-submarine detection and wartime nuclear research. He led the construction of a nuclear reactor in Montreal, Canada.

Paul Joseph Cohen (1934–2007) was an American mathematician. Cohen's early work focused on harmonic analysis, but he moved into mathematical logic after 1962. He invented the forcing method and showed that a series of independent results of axiomatic set theory, in particular the axiom of choice and the continuum hypothesis, were independent of the other axioms of ZF set theory. His breakthrough forcing method and independent results opened up many new theories of mathematics. He later tried to prove the Riemann conjecture, which was the only problem beyond his academic level. He was awarded the Bocher Memorial Medal of the American Mathematical Society in 1960, and the Fields Medal in 1966. In 1967, he was elected as a member of the National Academy of Sciences and won the National Medal of Science that year.

Alain Connes (1947–) is a French mathematician. His research interests include number theory, algebra, geometry, analysis, and mathematical physics. His outstanding achievement was to solve the problem of classification of Type II and Type III von Neumann algebras. More importantly, in the 1980s, he almost single-handedly founded a whole new field of study, non-commutative geometry, a subdiscipline that intersected analysis, algebra, topology, geometry and even quantum mechanics. Connes was awarded the Fields Medal in 1982, the Crafoord Prize in 2001, and the Gold Medal of CNRS (French National Centre for Scientific Research) in 2004. Connes is a member of the French Academy of Sciences and the National Academy of Sciences of the United States.

Richard Courant (1888–1972) was a German-American mathematician and member of the National Academy of Sciences. After working as an assistant and writing papers under David Hilbert, he received his doctorate at the University of Gottingen in 1910. He was a professor at the University of Münster and the University of Göttingen. He moved to the United States in 1934 and was a professor at New York University. He founded the New York University Institute for Mathematics and Mechanics, now known as the Courant Institute for Mathematical Sciences (CIMS). He is the author of *Methods of Mathematical Physics*, *Introduction to Calculus and Analysis*, and *What is Mathematics*, etc.

Francis Harry Compton Crick (1916–2004) was a British molecular biologist. He worked at the Cavendish Laboratory for Experimental Physics, Cambridge University from 1949 to 1953. During this period, he worked with James Dewey Watson to propose the famous theory of DNA double helix structure. Crick, Watson, and Maurice Hugh Frederick Wilkins won the 1962 Nobel Prize in Physiology or Medicine for proposing the theory of DNA double helix structure. Crick was a member of the Royal Society.

Pierre Curie (1859–1906) was a French physicist. He received his Ph.D. from the University of Paris, where he later became a professor. His early contributions were to the determination of the transition temperature of magnetic substances (Curie temperature), the establishment of Curie's law and the discovery of the piezoelectric phenomenon of crystals. Later, he worked with Madame Curie (Maria Skłodowska) on radioactive phenomena and discovered polonium and radium, two natural radioactive elements. For their contributions to the study of radioactivity, he shared the 1903 Nobel

Prize in Physics with Antoine Henri Becquerel and Madame Curie. He was elected to the French Academy of Sciences in 1905.

D

Julius Wilhelm Richard Dedekind (1831–1916) was a German mathematician. His book *Continuity and Irrational Numbers*, published in 1872, established a purely arithmetical basis for the real numbers by precisely explaining irrational numbers in terms of the division of rational numbers (now known as the Dedekind cut). In 1882, together with Heinrich Weber, he applied the ideal theory to algebraic function theory, laying the foundation of algebraic geometry in the general field and opening up the arithmetical direction of algebraic geometry. In 1899, he took the lead in the studying of lattice; he made a preliminary classification of finite lattice and became the founder of lattice theory. He also made important contributions to number theory, abstract algebra, and especially ring theory.

Pierre Deligne (1944–) is a Belgian mathematician. His major achievements include the proof of Weil's conjecture in 1973, co-constructing representations of algebraic groups using flat cohomology, assisting in the completion and development of Alexander Grothendieck's programme, establishing the theory of mixed Hodge structure, unification of Hodge theory, and advancing Galois representation, developing Shimura varieties theory, and collaborating with George Daniel Mostow on the monodromy group theorem of differential equations. Deligne received the Fields Medal in 1978 and the Crafoord Prize in 1988. He is one of the kindest mathematicians in the world and takes great delight in teaching his ideas to young people and non-experts alike.

Leonard Eugene Dickson (1874–1954) was an American mathematician who made important contributions to number theory and group theory. Dickson was hired as an associate professor of mathematics at the University of Texas at Austin in 1899. In 1900 he entered the University of Chicago, where he worked until 1939. As a prolific mathematician, Dickson pioneered extensive and in-depth research into finite field theory. He extended the theory of linear simultaneous equations by Joseph H. M. Wedderburn and Elie Cartan. One of his most notable studies was on the relationship between invariant theory and number theory. He used the analysis of the Russian mathematician Ivan Matveyevich Vinogradov to prove the ideal Waring theorem in his work on the theory of stacked numbers. Dickson was Professor Yang's father's thesis supervisor.

Paul Adrien Maurice Dirac (1902–1984) was a British theoretical physicist. While he was a graduate student, he developed a mathematical form of quantum mechanics, the q-number (non-commutative algebra) theory, becoming one of the founders of quantum mechanics. In 1928, he proposed the relativistic equations of motion of electrons (Dirac equations), which laid the foundation of relativistic quantum mechanics, profoundly gave a new physical meaning to vacuum, and correctly predicted the annihilation and production of positron pairs, leading to the recognition of the existence of antimatter. He was awarded the Nobel Prize in Physics in 1933.

Johann Peter Gustav Lejeune Dirichlet (1805–1859) was a German mathematician who greatly influenced the development of mathematics in Germany. In the field of analysis, he was one of the first mathematicians to advocate the rigorous method. In number theory, he was a propagator and popularizer of Gauss's thoughts, and gave a clear explanation of Gauss's difficult *Arithmetical Investigations*. In 1837, he introduced the Dirichlet series to prove that an arithmetic sequence contains infinitely many prime numbers. In 1846, he used the pigeonhole principle to clarify the structure of the abelian group of units in the algebraic number field. Both the Dirichlet type and the Dirichlet boundary value problems are named after Dirichlet.

Simon Donaldson (1957–) is a British mathematician whose research interests include the geometry and topology of four-dimensional differential manifolds. Using techniques developed from gauge field theory and particularly creative application of elliptic partial differential equations, he used Yang–Mills theory in the 1980s to find the series invariants of four-dimensional manifolds, and subsequently discovered that certain four-dimensional manifolds allow infinitely many differential structures. He was also one of the great and responsible mathematicians, willing to do something practical for the cause of mathematics.

Ronald George Douglas (1938–2018) was an American mathematician, best known for his work on operator theory and operator algebra. In 1977, he collaborated with Lawrence G. Brown and Peter A. Fillmore on a paper (BDF Theory) that introduced algebraic topology techniques into operator algebra theory. In addition to BDF theory, two influential theories bear his name: Douglas algebras and Cowen-Douglas operators, and he had been a prominent advocate of multivariable operator theory in recent decades.

Vladimir Drinfeld (1954–) is a Ukrainian mathematician. His research interests include number theory, algebraic geometry, quantum group theory, mathematical physics, etc. His major achievement is the first proof of Langlands conjecture for function fields in the case n = 2. He is one of the founders of the theory of quantum groups, which is closely related to mathematical physics. He named a class of infinite dimensional Hopf algebras after Dr. Yang Chen-Ning called Yangian (Young's symmetry relation), which is used as a tool for solving the quantum Yang–Baxter equation. He was awarded the Fields Medal in 1990.

Freeman Dyson (1923–2020) was a British-born American physicist and educator best known for his speculative work on extraterrestrial civilizations. He was the author of *Disturbing the Universe* (1979), *Weapons and Hope* (1984), *Origins of Life* (1985), *Infinite in All Directions* (1988), *Imagined Worlds* (1998), *The Sun, the Genome, and the Internet* (1999), among others. Dyson was a Fellow of the Royal Society and a member of the National Academy of Sciences. He was awarded the Wolf Prize in Physics in 1981, the Lewis-Thomas Prize in 1996, the Templeton Prize for Progress in Religion in 2000, and the Henri Poincare Prize at the International Congress of Mathematical Physics in 2012. He became famous before graduating, so he did not receive a Ph.D. He probably should have won the Nobel Prize, but he didn't.

F

Ludwig D. Faddeev (1934–2017) was a Russian physicist. Since the 1970s, Faddeev has been working on the quantum theory of solitons, which opened up new paths for quantum field theory and led to the new concept of quantum groups. He was elected to the Russian Academy of Sciences in 1976. He was also a member of the French Academy of Sciences, the National Academy of Sciences, and a fellow of the Royal Society. He was awarded the Demidov Prize of the Russian Academy of Sciences in 2002, the Henri Poincare Prize in 2006, and the Shaw Prize in 2008.

Michael Faraday (1791–1867) was a British physicist and chemist. He became a fellow of the Royal Society in 1824. In 1835, the direction of induced current was experimentally determined, thus laying the foundation for the development of the whole electromagnetism. In 1850, he put forward a new idea about the force in space, which was another unique idea that shook the scientific community the most since Newton.

Pierre de Fermat (1601–1665) was a French mathematician who made significant contributions to number theory, solving geometry, calculus, probability theory, and the variational principle. Fermat proved or put forward many propositions in number theory. The most famous is Fermat's Last Theorem. He was a pioneer of calculus, but he also put forward the Fermat principle of optics, which gave great inspiration to the later study of variational method. He was the author of *Ad Locos Planos et soli-dos isagoge*.

Enrico Fermi (1901–1954) was an Italian-American physicist. In 1929, he became a member of the Royal Society of Italy, which existed for a short time, and a foreign member of the Royal Society in 1950. The United States Atomic Energy Commission established the Fermi Prize, the first of which was awarded to him in 1954. Fermi made important contributions to statistical physics, atomic physics, nuclear physics, and particle physics. Fermi was awarded the Nobel Prize in Physics in 1938.

Richard Feynman (1918–1988) was an American physicist. He received his Ph.D. from Princeton University in 1942 and was elected to the National Academy of Sciences in 1954. Feynman developed the method of expressing quantum amplitudes by path integrals in the 1940s, and proposed a new theoretical form, calculation method, and renormalization method in the field of quantum electrodynamics in 1948. For this contribution, he shared the 1965 Nobel Prize in Physics with Julian Schwinger and Tomonaga Shinichiro. Feynman diagram, Feynman amplitude, Feynman propagator and Feynman rule in quantum field theory are all named after him. He was the author of several books, including *The Feynman Lectures on Physics* and *Surely You're Joking, Mr. Feynman!*

Michael Hartley Freedman (1951–) is an American mathematician who received his doctorate from Princeton University in 1973. Freedman mainly studied topology. He proved Poincare's conjecture for the four-dimensional manifold, and thus made a major breakthrough in the topology of four-dimensional manifolds. He not only completed the characterization of the four-dimensional sphere, but also provided a complete classification of four-dimensional compact simply connected topological manifolds, which gave rise to a series of previously unknown four-dimensional manifolds and unknown homeomorphisms between known manifolds. After topology, he began working on quantum computers at Microsoft. Freedman was elected to the National Academy of Sciences in 1984. He was awarded the Fields Medal in 1986, the Veblen Medal of the American Mathematical Society in the same year, and the National Medal of Science in 1987.

G

Johann Carl Friedrich Gauss (1777–1855) was a German mathematician, astronomer, and physicist. He made a series of seminal contributions in number theory, algebra, non-Euclidean geometry, differential geometry, hypergeometric series, complex variable function theory, and elliptic function theory. His book *Disquisitiones Arithmeticae* was a classic in the history of mathematics, opening up a new era in the study of number theory, and the book is still highly readable today. Gauss loved his first wife very much, and the words he wrote about her deeply touched the hearts of many.

Israel M. Gelfand (1913–2009) was a Ukrainian mathematician and biologist. He was elected a corresponding member of the Soviet Academy of Sciences in 1953 and received the Wolf Prize in Mathematics in 1978. Gelfand established normed ring theory, or commutative Banach algebra theory. Using algebraic methods, he introduced the maximal ideal subring space and gave the concept of the representation of elements on it (Gelfand representation), which led to the further study of linear operator spectrum theory. He was a very visionary mathematician who influenced many disciplines and many people, for example, his famous Moscow seminars trained a large number of talents.

Murray Gell-Mann (1929–2019) was an American theoretical physicist. He conducted research on quantum field theory, nuclear physics, and particle physics. He was elected a member of the National Academy of Sciences in 1960 and won the Nobel Prize in Physics in 1969. Gell-Mann was best known for discovering the eightfold way in physics, the organization of a class of subatomic particles called hadrons, which led to the development of the quark model. The name eightfold way is inspired by the Eight Sacred Paths of Buddhism.

Phillip A. Griffiths (1938–) is an American mathematician. He pioneered, with collaborators, the variation theory of Hodge structures, which has played a central role in many aspects of algebraic geometry, as well as in its application to modern theoretical physics. In addition to algebraic geometry, he has made contributions to differential and integral geometry, the theory of geometric functions, and the geometry of partial differential equations. He won the Wolf Prize in Mathematics in 2008.

David Jonathan Gross (1941–) is an American theoretical physicist. In 1973, Frank Wilczek and he published a paper proposing asymptotic freedom theory of strong interactions in particle physics. At the same time, H. David Politzer independently proposed a similar theory. The three shared the 2004 Nobel Prize in Physics for their work. Gross has made pioneering contributions to gauge field theory, particle physics, and superstring theory. Gross is one of the founders of hybrid string theory, which became one of the most important works that led to the first revolutionary advances in string theory. Gross was elected an academician of the National Academy of Sciences in 1986 and a foreign academician of the Chinese Academy of Sciences in 2011.

Alexander Grothendieck (1928–2014), a stateless mathematician living in France, was born in Berlin, Germany. Grothendieck's main work in the early stage was functional analysis, introducing the theory of nuclear space and the tool of tensor product. This work was summarized in the book *Topological Tensor Product and Nuclear Space* published in 1955. Subsequently, he conducted systematic research on homology algebra, especially Abelian category theory, and used it to construct new basic structures for algebraic geometry, thus completely changing the face of algebraic geometry and related disciplines. He was awarded the Fields Medal in 1966 and the Crafoord Prize of the Royal Swedish Academy of Sciences in 1988, but he declined to accept the award. Dynamic and insightful, Grothendieck was perhaps one of the most idealized and dramatic mathematicians of the 20th century, and his work would profoundly influence the development of mathematics.

H

Jacques Salomon Hadamard (1865–1963) was a French mathematician. He was elected a member of the French Academy of Sciences in 1912. His doctoral thesis (1892) was the first to introduce set theory into the theory of functions of complex variables, which more simply reproved Cauchy's results on the radius of convergence. He discovered the famous three circles theorem in the study of maximal modules of functions. He was the first to prove the prime number theorem in number theory, and his contributions in the real domain are in the qualitative theory of ordinary differential equations, functional analysis, linear second-order partial differential equations with definite solutions, and fluid mechanics. Hadamard was a generalist who had knowledge of all the fields of mathematics of his day. He was the author of *A Course in Variational Calculus*, *A Course in the Propagation of Waves*, and *The Psychology of Inventions in Mathematics*.

William Rowan Hamilton (1805–1865) was a mathematician and physicist. He was a professor at University College Dublin, director of the Observatory of the University College Dublin, and president of the Royal Irish Academy. He made important contributions to the development of analytical mechanics. In 1834, he established Hamilton's principle and first proposed the concept of quaternions, thus freeing algebra from the commutative postulate of multiplication. He also explained the phenomenon of conical refraction and contributed to the establishment of modern vector analysis methods.

Godfrey Harold Hardy (1877–1947) was an English mathematician. Hardy's contributions to number theory, including the Diophantine approximation, additive number theory, prime distribution theory and Riemann ζ function, trigonometric series theory in harmonic analysis, summation of divergent series and Tauberian theorem, inequalities, integral transformations, and integral equations had a profound impact on the development of analysis. The theory of H^p Spaces (Hardy Spaces), which bears his surname, remained a very active field in mathematics. Others, such as the Hardy-Littlewood maximum function and Hardy's inequality, were often cited. Hardy wrote many books on mathematics, but his most popular work was not a book on mathematics; it was *A Mathematician's Apology*.

Harish-Chandra (1923–1983), born in Kanpur, British India (now Uttar Pradesh, India), did much pioneering work on the representation theory of Lie groups. His main interest was in the study of representations of infinite dimensions, problems that first emerged in quantum mechanics in the 1930s when it became necessary to disentangle the effects of symmetry on the motion of particles and waves. The theory of representations constructed by Harish-Chandra had an impact on several fields such as geometry and number theory. He became a fellow of the Royal Society in 1973 and was awarded the Ramanujan Medal of the Indian Academy of Sciences in 1974. Chandra was methodical and wrote handouts that could be printed straight into a book.

Werner Karl Heisenberg (1901–1976) was a German physicist and one of the founders of quantum mechanics. In 1925, he proposed that the unobtainable mechanical properties of microscopic particles, such as position and momentum, should be expressed by the observable frequencies and intensities of their spectra after certain operations (matrix algebraic rules). He then collaborated with Max Born to establish matrix mechanics, which played a precursor role in the establishment of quantum mechanics. In 1927, he proposed the uncertainty principle. For his contributions, he was awarded the Nobel Prize in Physics in 1932.

David Hilbert (1862–1943), a German mathematician and one of the most influential mathematicians of the late 19th and early 20th centuries, made significant contributions to many branches of mathematics, including: invariant theory, algebraic number theory, analytic number theory, variational method, foundations of geometry. For example, he solved the long-pending Waring's problem. Professor Yang Chen-Ning's father and his thesis advisor also worked on other versions of the Waring's problem. Hua Loo-Keng's first paper also discussed the Waring's problem. For physicists, Hilbert's most important contribution may have been the theory of Hilbert Spaces, one of the foundations of functional analysis. A series of problems he presented at the International Congress of Mathematicians in 1900 (Hilbert's 23 problems) guided much of the mathematical research of the 20th century. Hilbert and his students made important contributions to forming the mathematical foundations of quantum mechanics and general relativity. He was also one of the founding fathers of proof theory, mathematical logic, and the distinction between mathematics and metamathematics. Hilbert took over Klein's career and enabled the University of Gottingen to maintain its position as a world center of mathematics.

Gerardus 't Hooft (1946–) is a Dutch theoretical physicist and member of the Netherlands Academy of Sciences. Hooft assisted Professor Martinus J. G. Veltman in studying renormalization methods of non-Abelian gauge theory in the autumn of 1970. For their work on renormalization gauge theory, which formulated the quantum structure of the electroweak interaction, the unified theory of electroweak was widely accepted, and Hoft and Weltman shared the 1999 Nobel Prize in Physics.

J

Carl Gustav Jacob Jacobi (1804–1851) was a Prussian mathematician widely regarded as one of the most distinguished mathematicians in history. One of his most distinguished achievements was the theory of elliptic functions and their relation to the elliptic Θ function. His work on differential equations and classical mechanics, especially the Hamilton-Jacobi equation, made fundamental contributions to the field. His academic controversy with Abel contributed greatly to the rapid development of the theory of elliptic functions. The basic Jacobian identity in Lie algebra theory was named after him.

Camille Jordan (1838–1922) was a French mathematician who worked at the École Polytechnique and L'Institut de France. His founding contribution to group theory was his *Treatise on Substitutions and Algebraic Equations*, a fundamental research treatise that laid a firm foundation for both Galois and group theory. This book had an important influence not only on the establishment of Galois's theory, but also on the work of Felix Klein and Marius Sophus Lie in the use of groups in geometry and in Lie theory. Jordan's highly influential *Tutorial in Analysis* was very important. For example, it contained the famous Jordan curve theorem in topology.

Brian David Josephson (1940–) is a British physicist. He won the 1973 Nobel Prize in Physics for predicting superconducting currents in tunnels. He was awarded the F. London Prize, Van der Pol Medal, Cresson Medal, Hughes Medal, Holweck Medal.

K

Mark Kac (1914–1984) was a Polish-American mathematician. His main interest was in probability theory. In 1959, he published a classic article "Statistical Independence in Probability Theory, Analysis, and Number Theory." His most famous work was the Feynman-Kac formula, which linked parabolic partial differential equations to stochastic processes.

Charles Kuen Kao (1933–2018) was a Chinese-American physicist. Born in Shanghai, China, he received his Ph.D. from University College London (UCL) in 1965. He was the president of the Chinese University of Hong Kong from 1987 to 1996. Gao was awarded 28 patents in electromagnetic waveguides, ceramic science, and optical fiber manufacturing. He was elected an academician of the US National Academy of Engineering in 1990, a member of the Academia Sinica in Chinese Taiwan in 1992, a foreign academician of the Chinese Academy of Sciences in 1996, and a fellow of the Royal Society in 1997. He was awarded the Nobel Prize in Physics in 2009 and the Grand Bauhinia Medal of the Hong Kong Special Administrative Region of China in 2010.

David Kazhdan (1946–) was born in Moscow, emigrated to the United States, and served at Harvard University (1975–2002) before moving to Israel as a professor at the Hebrew University of Jerusalem. He is a member of the National Academy of Sciences and a member of the American Academy of Arts and Sciences. He is concerned with the different forms of symmetry that occur in nature and mathematics, with an emphasis on algebraic aspects of analysis and geometry, as well as representation theory. His ideas have influenced work in such areas as automatic representation, differential geometry, differential equations, representation of finite groups, and number theory. He is best known for his work on the Kazhdan-Lusztig conjectures, which shows that the fundamental invariance of most representation theories is actually the invariance of intersecting cohomology. Another famous and influential work is his paper on the Kazhdan property (T).

Johannes Kepler (1571–1630) was a German astronomer and physicist. In 1600, at the invitation of Tycho Brahe, Kepler went to work at the Prague Observatory and continued Tycho Brahe's business after his death. He summarized Tycho Brahe's observations, discovered that planets move in elliptical orbits, and proposed the three laws of planetary motion, which laid the foundation for Newton's discovery of the law of universal gravitation. Kepler also compiled star catalogs and discovered the approximate law of atmospheric refraction. He wrote *Mysterium Cosmographicum, Astronomiae Pars Optica, Harmonices Mundi* and *Epitome astronomiae Copernicanae.*

Wilhelm Karl Joseph Killing (1847–1923) was a German mathematician who made important contributions to Lie algebras, Lie groups, and non-Euclidean geometry. His paper on the classification of complex simple Lie algebras was sometimes called "the greatest mathematical paper ever written."

Felix Klein (1849–1925) was a German mathematician. In 1871, Klein published his research on the synthesis of non-Euclidean geometry from the idea of projective geometry. In 1872, when he was appointed professor at the University of Erlangen, he further developed these ideas by writing a pamphlet as part of his inaugural address, which became known as the Erlangen Program. Klein lost the academic battle with Poincare, which destroyed his research career, and he turned to teaching and writing. His major works included *Lectures on the Theory of Automorphic Functions*, *Elementary Mathematics from an Advanced Standpoint* and *Vorlesungen über die Entwicklung der Mathematik im 19*. With a strong personality, he single-handedly built the University of Gottingen into a world center of mathematics in the early 20th century.

Maxim Kontsevich (1964–) is a Russian mathematician. Kontsevich's research work is mainly concerned with mathematical structures in modern theoretical physics. He independently formulated two-dimensional conformal field theory. In his doctoral thesis, he proved the Witten conjecture, in which the Feynman diagram technique was first used in mathematics. Later, he proved the main results of Vassiliev's invariant theory by field theory method. He discovered the profound relationship between calculus operators, cohomology of Lie algebras, Feynman graphs, and topology, and studied theories such as quantum cohomology and mirror symmetry from a new perspective. He also made contributions to the quantization of Poisson manifolds, as well as to counting geometry. In 1998, he received the Fields Medal.

Polykarp Kusch (1911–1993) was an American physicist. He received his Ph.D. from the University of Illinois in 1936. He was elected member of the National Academy of Sciences in 1956. He began his research with Isidor Isaac Rabi in atomic, molecular and nuclear physics using molecular beam magnetic resonance (MMR). In 1947, Henry Michael Foley and he used the molecular beam magnetic resonance method to accurately determine the intrinsic magnetic moment of the electron, which was not exactly equal to one Bohr magnet. He also accurately determined the magnetic moments of electrons in various atoms and molecules and the properties of nuclei, which advanced the development of quantum electrodynamics. In 1955, he was awarded the Nobel Prize in Physics together with Willis Eugene Lamb.

L

Joseph-Louis Lagrange (1736–1813) was a French mathematician, astronomer, and mechanic. Lagrange's *Theorie des fonctions analytiques* (1797) and *Lecons sur le calcul des fonctions* (1804) were among the earliest teaching texts on analytic functions. Together with Laplace, he established a rigorous system of classical celestial mechanics, further developed and perfected Euler's variational method, and solved the mutual perturbation of Jupiter and Saturn. From 1764 to 1778, he was awarded the French Academy of Sciences Prize five times for his research on celestial mechanics such as the translation of the moon. Lagrange's work on algebraic equations was the beginning of Galois' work, and he made fundamental contributions to number theory.

Willis Eugene Lamb (1913–2008) was an American physicist who was elected to the United States National Academy of Sciences in 1954. Lamb's main contribution was the discovery of the Lamb shift. The Lamb shift experiment, together with the experiments on the anomalous magnetic moments of electrons and Muons, form the three main experimental pillars of quantum electrodynamics. Anomalous electron magnetic moments were discovered by Polykarp Kusch. For this work, Lamb and Kusch shared the 1955 Nobel Prize in Physics.

Lev Davidovich Landau (1908–1968) was a Soviet theoretical physicist who founded the Soviet School of Theoretical Physics. The terms that bear his surname are Landau's magnetism, Landau levels in solid state physics, Landau damping in plasma physics, Landau energy spectrum in low-temperature physics, and Landau shunt effects in high-energy physics. He was awarded the Nobel Prize in Physics in 1962 for his work on condensed matter theory, particularly liquid helium theory. With his student Evgeny Lifshitz, Landau published the famous *Course of Theoretical Physics*, a ten-volume series covering the entire field of theoretical physics. He was a very strict and strong man, both at work and in his love life. For example, he said that *Course of Theoretical Physics* is required by every experimental physicist to understand the background of theoretical physics.

Robert Langlands (1936–) is a Canadian mathematician. His most important contribution is the development of the Langlands Program. The program is a theoretical system that combines number theory, group representation theory, non-commutative harmonic analysis and automorphic formalism. It extends Abelian class field theory, Hecke theory, automorphic function theory, and representation theory of reducible groups. The program contains a large number of conjectures (with a few special cases proved), including how Langlands and his collaborators proved the GL(2) case and the Artin's conjecture of the octahedron, which became the starting point for solving Fermat's last Theorem. Langlands was awarded the Kohl Prize of the American Mathematical Society (1982), the Inaugural Prize in Mathematics of the National Academy of Sciences (1988), the Wolf Prize in Mathematics (1995–1996), and the Shaw Prize (2007). He was elected a fellow of the Royal Society of Canada in 1972 and a fellow of the Royal Society in 1981.

Pierre-Simon Laplace (1749–1827) was a French mathematician, astronomer, and physicist. Laplace studied a wide range of subjects, including astronomy, mathematics, physics, and chemistry. Most of his work was contained in three concluding works: *Universe System Theory* (1796), *Celestial Mechanics* (1799–1825), and *The Analytical Theory of Probability* (1812). Because of his important and extensive scientific achievements, Laplace was called the French Newton.

Marius Sophus Lie (1842–1899), a Norwegian mathematician, was the founder of continuous transformation group theory. This theory was fundamentally important to the development of differential equations and differential geometry in the 20th century and has evolved into the independent theory of Lie groups and Lie algebras. He was the author of *Transformation*

Groups, Differential Equations, Continuous Groups, etc. He was a close friend of Klein. Both Lie and Klein benefited from their friendship, but for a long time there was a deep estrangement between them, and the relationship fell apart. It was not until a short time before Lie's death that their relationship was finally repaired, and they were happily reunited.

Evgeny Lifshitz (1915–1985) was a Soviet theoretical physicist and member of the Soviet Academy of Sciences. In 1935, Lifshitz and his mentor L. D. Landau developed an equation describing the dynamics of the magnetic moments of ferromagnets in the presence of external fields and spin-orbit interactions (the Landau–Lifshitz equation), which was the basis of modern ferromagnetism theory. Together, they developed the theory of ferromagnetic resonance and the complete theory of domain structure of ferromagnets. *Course of Theoretical Physics* by Lifshitz and Landau was a world-renowned work on theoretical physics and a landmark work reflecting the transition from classical physics to modern physics. He was awarded the Lenin Prize in 1962.

John Edensor Littlewood (1885–1977) was a British mathematician. Much of his work was in collaboration with G. H. Hardy. His work in number theory included the theory of the distribution of prime numbers, the Waring's problem, the Riemann ζ function, trigonometric series theory in harmonic analysis, summation of divergent series and Tauber's theorem, inequalities, single leaf function and nonlinear differential equation, which have made important contributions. Since 1931, he worked with Raymond E. A. C. Paley on Fourier series and power series, establishing the Littlewood-Paley theory. The Hardy-Littlewood maximum function was also frequently cited. Littlewood's famous *A Mathematician's Miscellany* contained his autobiography and many anecdotes, which was well worth reading.

Hendrik Antoon Lorentz (1853–1928) was a Dutch physicist, with a Ph.D. from Leiden University, where he later became a professor. He was a fellow of the Royal Society. He founded classical electron theory and made important contributions to classical electromagnetic theory. He calculated the force on electrons in the electromagnetic field (Lorentz force) and predicted the normal Zeeman effect. He was awarded the 1902 Nobel Prize in Physics together with Dutch physicist Pieter Zeeman for his work on the effect of magnetism on radiation.

M

Robert MacPherson (1944–) is an American mathematician at the Institute for Advanced Study and Princeton University. He is best known for his work with Mark Goresky on the theory of intersection homology. Following his Chern class definition of singular varieties, in 1983, he gave a general assembly presentation at the International Congress of Mathematicians in Warsaw. In 1992, MacPherson was awarded the mathematics award of the National Academy of Sciences. In 2002, he and Goresky were awarded the Steele Prize by the American Mathematical Society for groundbreaking research contributions. His collaboration with Goresky was fruitful and continued for many years.

Yuri Ivanovich Manin (1937–2023) was a Russian mathematician. He worked on algebraic geometry and Diophantine geometry, and carried out interpretive work in the fields of mathematical logic and theoretical physics. Manin was one of the first to propose the idea of a quantum computer and is the author of *Computable and Uncomputable*. Manin was one of the best students to come out of Moscow University in its golden years.

Gregori Aleksandrovich Margulis (1946–) is a Russian-American mathematician and a winner of the Fields Medal, the Wolf Medal in Mathematics, and the Abel Prize. He is a great mathematician, deeply thoughtful, but seemingly slow to grasp the work and interpretations of others. David Kazhdan and he co-wrote a paper in their junior year of college that solved the conjecture of Atle Selberg (a Fields Medalist), and the paper was written by Armand Borel. It was presented by Borel at the prestigious Bourbaki Seminars.

James Clerk Maxwell (1831–1879) was a Scottish mathematical physicist. His greatest achievement was the formulation of Maxwell's equations, which classified electricity, magnetism, and light into electromagnetic field phenomena, achieving the second unification of physics since Isaac Newton. In 1864, he published a paper entitled "A Dynamical Theory of the Electromagnetic Field." The Maxwell Institute of mathematics is now established in Edinburgh, where Maxwell was born.

Lise Meitner (1878–1968) was an Austrian-Swedish atomic physicist. She was the first person to theorize the discovery of nuclear fission by Otto Hahn in 1938. She was perhaps the most deserving of a Nobel Prize for a prominent female scientist, yet she failed to win one.

John Willard Milnor (1931–) is an American mathematician. His main contribution lies in differential topology, K-theory, and dynamical systems and their writings. He was the recipient of the 1962 Fields Medal, the 1989 Wolf Prize in Mathematics, and the 2011 Abel Prize. While still a college student, he became well known for solving a famous unsolved problem. Legend had it that he mistook the problem for a homework assignment and then managed to solve it. He shocked the mathematical world with his construction of a strange sphere in seven dimensions, and for the first time proposed an unexpected distinction between a topological flow and a differential manifold. He was an excellent author of mathematical books, many of which have become contemporary classics, such as *Characteristic Classes*, *Topology from the Differentiable Viewpoint*, and *Morse Theory*. He was also a great speaker.

Robert Laurence Mills (1927–1999) was an American physicist. His main contribution was the Yang–Mills theory, in which he cooperated with Yang Chen-Ning.

Hermann Minkowski (1864–1909) was a German Jewish mathematician, the creator of the four-dimensional space-time theory, and the teacher of the famous physicist Albert Einstein. He was a child prodigy. At the age of 19, he caused a sensation in British mathematics when he shared the prestigious mathematical prize of the Paris Academy of Sciences with Henry Smith (1826–1883), the most outstanding and accomplished English mathematician of the time. In addition to his work on general relativity, he founded a very original branch of number theory: the geometry of numbers. He was also a close friend of David Hilbert, advising him on Hilbert's famous list of 23 problems.

Jürgen K. Moser (1928–1999) was an American mathematician born in Germany. His research interests included Hamiltonian dynamical systems and partial differential equations. His most important work was his contribution to KAM theory.

David Bryant Mumford (1937–) is an American mathematician who also studied pattern science in computer science. He developed the theory of geometric invariants and used it to construct the modular space of algebraic curves, a problem posed by Riemann. These conclusions were published in Mumford's book *Geometric Invariants Theory*. He was awarded the Fields Medal in 1974, the Shaw Prize in 2006, and the Wolf Prize in Mathematics in 2008. After the death of his first wife, he left mathematics, as well as Harvard, to pursue computer vision research at Brown University. Mumford later became interested in the history of mathematics in ancient China as well as in India.

N

John Forbes Nash, Jr. (1928–2015) was an American mathematician who worked on game theory, differential geometry, and partial differential equations. He developed the concept of the Nash equilibrium, which became an important breakthrough in game theory. In 1994, he shared the Nobel Prize in Economics with two other game theorists, John C. Harsanyi (1920–2000) and Reinhard Selten (1930–2016). For his work on nonlinear partial differential equations, he shared the 2015 Abel Prize with Louis Nirenberg (1925–2020). Although he was best known for his work in game theory, his work on pure mathematics, such as the Nash embedding theorem, was one of the most original works in the 20th century. Nash's life was full of twists and turns and sadness, as depicted in his famous biography and Hollywood film, as was his unexpected death (which, of course, was not mentioned in the book or the film). Feeling rich after winning the Abel Award, Nash decided to take a taxi home from the airport on his return trip, where he was killed along with his wife in a car accident on the highway. He and his wife had had a difficult couple of years, and in a way, they were lucky to die together.

John von Neumann (1903–1957) was an American mathematician who was born in Hungary. He was the founder of theoretical computer science and game theory, and made significant contributions to many mathematical fields such as functional analysis, ergodic theory, geometry, topology, and numerical analysis, as well as computer science, quantum mechanics and economics. He had a strong sense of perception and a good memory. In his schooldays, his performance in class often intimidated other great mathematicians, such as George Pólya.

Emmy Noether (1882–1935) was a German mathematician and a prominent figure in the fields of abstract algebra and theoretical physics. Pavel S. Aleksandrov, Albert Einstein, Jean Alexandre Eugène Dieudonné, Hermann Weyl and Norbert Wiener, among others, hailed Noether as the most prominent female mathematician in history. Her work in the field of mathematics included rings, fields, and algebras over fields; and in physics, her proof of Noether's theorem revealed a close relationship between symmetry and conservation laws. She was not a methodical speaker, but her talk was inspiring. She surrounded herself with a large group of brilliant young mathematicians, often referred to as the Noether boys. Her father was also a famous mathematician.

O

Lars Onsager (1903–1976) was born in Oslo, Norway, moved to the United States in 1928, and was elected to the National Academy of Sciences in 1947. In 1925, he proposed the Onsager Limit Law for electrolyte theory. In 1931, he proved the Onsager reciprocity relation that now bears his name, one of the principal theories of the thermodynamics of linear irreversible processes. This relation and his later principle of minimum dissipation of energy in a fixed state laid the foundation for the quantitative theory of irreversible thermodynamics and its application. A paper he published in 1949 laid a solid statistical mechanical foundation for liquid crystal theory; and from 1951 to 1952, he proposed a theory of the diamagnetism of metals. He was awarded the Nobel Prize in Chemistry in 1968.

Julius Robert Oppenheimer (1904–1967) was an American physicist known as the father of the atomic bomb for his work on the Manhattan Project. He was director of the Institute for Advanced Study at Princeton from 1947 to 1966 and was closely associated with Dr. Yang Chen-Ning.

P

Roger Penrose (1931–) is an English mathematical physicist. For his significant contributions to general relativity and cosmology, he won the 2020 Nobel Prize in Physics for his discovery that the formation of black holes is a definitive prediction of general relativity. To the general public, he is perhaps best known for his Penrose tiling, which can be used in bathrooms to make them more beautiful and secure.

Grigori Perelman (1966–) is a Russian mathematician who was born in Leningrad, USSR (now St. Petersburg), and an expert on Ricci flow, who made a decisive contribution to the proof of Poincaré conjecture. His eccentric personality and behavior may have provided more evidence for the popular myth that mathematicians are a little weird.

Max Karl Ernst Ludwig Planck (1858–1947) was a German physicist and founder of quantum mechanics. He was awarded the 1918 Nobel Prize in Physics for his discovery of energy quanta. Planck's constant, named after him, was used in 2019 to redefine the fundamental unit.

Jules Henri Poincaré (1854–1912) was a French mathematician, theoretical scientist and philosopher of science. He made many creative and foundational contributions to mathematics, mathematical physics, and celestial mechanics. His Poincaré conjecture was one of the most famous problems in mathematics, which was only a small part of the algebraic topological theory he constructed. In his work on the three-body problem, Poincaré became the first person to discover a chaotic deterministic system and laid the foundation for modern chaos theory. His work on rational points on elliptic curves led to one of the most famous problems in number theory, Birch and Swinnerton-Dyer conjecture, one of seven Millennium Prize Problems selected by a committee at the Clay Mathematics Institute in the United States. In addition to his many profound papers, Poincaré also published three famous popular science works that attracted generations of well-educated public.

George Pólya (1887–1985) was a Hungarian-American mathematician who was elected to the National Academy of Sciences in 1976. His mathematical research covered many fields, such as functions of complex variables, probability theory, number theory, applied mathematics, mathematical analysis, and combinatorics. The Pólya enumeration theorem, proposed in 1937, was an important tool in modern combinatorial mathematics. He was engaged in mathematics teaching for a long time and conducted in-depth research on the general laws of mathematical thinking. He wrote several books telling students how to learn and do research, such as the famous *Problems and Theorems in Analysis* and *How to Solve It,* which are still highly recommended to ambitious students.

R

Isidor Isaac Rabi (1898–1988) was an American physicist who was elected to the National Academy of Sciences in 1940 and was president of the American Physical Society in 1950. Rabi developed Stern's molecular beam method and, in 1938, used molecular beam and atomic beam magnetic resonance to study the emission spectrum, accurately determining the magnetic moments of atoms and molecules and laying the foundation of radio frequency spectroscopy. He participated in radar research from 1940 to 1945. He was a member of the General Advisory Committee of the United States Atomic Energy Commission from 1946 to 1956 (chairman of which from 1952 to 1956). He was also a founding member of the Brookhaven National Laboratory. He was awarded the Nobel Prize in Physics in 1944.

Srinivasa Ramanujan (1887–1920) was an Indian mathematician. His research interests included number theory, hypergeometric series, elliptic function theory, and continued fractions. His work was largely based on intuition and often lacked rigorous proof, but his results were mostly correct, for reasons that had never been fully explained. In 1985, Bruce Carl Berndt, an American mathematician, collated and proved Ramanujan's results in his notebooks and published a three-volume collection, *Ramanujan's Notebooks.*

Bernhard Riemann (1826–1866) was a German mathematician and mathematical physicist. Riemann did not write much, but his influence was extremely profound. In his doctoral thesis in 1851, he demonstrated the necessary and sufficient conditions for derivability of complex variable functions (now commonly known as Cauchy–Riemann equations) and expounded the Riemann mapping theorem with the help of Dirichlet principle, which became the basis of the geometric theory of functions. In 1853, he defined the Riemann integral and studied the convergence criterion of trigonometric series. His inaugural address established the concept of Riemannian space in 1854. The paper in 1857 on the Abelian function led to the concept of Riemannian surface and the formulation of the Riemann-Roch theorem, which was later supplemented by G. Roch. In 1858, he published a paper on the distribution of prime numbers and put forward the Riemannian conjecture. In addition, he made significant contributions to partial differential equations and their applications in physics.

Julia Hall Bowman Robinson (1919–1985) was an American mathematician. Robinson's reputation was associated with her contribution to the solution of Hilbert's tenth problem. The problem was eventually solved by her and Hilary Putnam, Yuri Matiyasevich and Martin Davis, who showed that no such algorithm existed. Robinson became the first elected female member of the National Academy of Sciences in 1975. In the 1980s, she became the first female president of the American Mathematical Society. Her sister was the author of Hilbert's famous autobiography, and Constance Reid was influenced by her.

Wilhelm Conrad Röntgen (1845–1923) was a German physicist with a Ph.D. from the University of Zurich, a professor and the president of the University of Würzburg, and the director of the Institute of Physics of the University of Munich. He was also the corresponding member of the Berlin Academy of Sciences and the Munich Academy of Sciences. In 1895, X-rays were discovered and studied in depth, so they are also called Röntgen rays. In addition, he discovered that a magnetic field can be generated when the medium was rotated in a fixed parallel-plate capacitor that is charged. He was awarded the first Nobel Prize in Physics in 1901.

S

Richard Melvin Schoen (1950–) is an American mathematician. He studied differential geometry and received his Ph.D. from Stanford University in 1977 under the supervision of Leon Simon and Shing-Tung Yau. He was Yau's best student and chief collaborator. He won the Wolf Prize in Mathematics in 2017 and gave general conference presentations twice at the International Congress of Mathematicians.

Erwin Schrödinger (1887–1961) was an Austrian theoretical physicist and one of the founders of quantum mechanics. In 1926, Schrödinger proposed the Schrödinger equation, which laid a solid foundation for quantum mechanics, and proposed the "Schrödinger's cat" thought experiment, which attempted to prove the incompleteness of quantum mechanics under macroscopic conditions. In 1933, he shared the Nobel Prize in Physics with British physicist Paul Adrien Maurice Dirac (1902–1984) for their discovery of new forms useful in the theory of the atom. Schrödinger had many mistresses, especially young ones. When he was vacationing with his mistresses, he discovered the Schrödinger equation. Schrödinger believed that love for women was a part of his life that could not be isolated from his work. There is an academic institution named after him in Vienna, and his portrait appears on an Austrian banknote.

Laurent Schwartz (1915–2002), a French mathematician and graduate of the École Normale Supérieure in Paris, won the Fields Medal in 1950 for his theory of distribution, which gave clear mathematical definitions such as the Dirac δ function.

Julian Schwinger (1918–1994) was an American theoretical physicist and one of the founders of quantum electrodynamics, who shared the 1965 Nobel Prize in Physics with Richard Feynman and Shinichirō Tomonaga. Schwinger was known among physicists as a master of hard arithmetic, with a knack for long, difficult pen calculations.

Atle Selberg (1917–2007) was a Norwegian mathematician. He was known for his work in analytic number theory, as well as the theory of automorphic forms, especially its introduction into spectral theory, where he formulated the Selberg trace formula. He also investigated the rigidity of discrete subgroups of semisimple Lie groups, which inspired such fundamental work as strong rigidity by George Daniel Mostow, superrigidity by Gregori A. Margulis, and arithmetical properties of lattices. Selberg won the 1950 Fields Medal and the 1986 Wolf Prize in Mathematics.

Jean-Pierre Serre (1926–) is a French mathematician whose main contributions are to the fields of topology, algebraic geometry and number theory. He was the recipient of many prizes in mathematics, including the Fields Medal in 1954 and the Abel Prize in 2003. He remains the youngest mathematician to win the Fields Medal to date. Although in his nineties, he is still very active. He enjoys playing table tennis and always wants to win in a fair way. He loves rock climbing, and he loves good wine—probably as much as he likes good math! He wrote many beautiful and influential books.

Goro Shimura (1930–2019) was a Japanese mathematician and an honorary professor at Princeton University, specializing in number theory, automorphic forms, and arithmetic geometry. His contributions to mathematics include Shimura varieties and complex multiplication of Abelian varieties. His work with Yutaka Taniyama on the Taniyama-Shimura conjecture was a key part of the solution to Fermat's Last Theorem. As one of Japan's most famous mathematicians, he grew up in the difficult years during and after World War II. He also had an in-depth knowledge of Chinese literature and philosophy.

Karl Manne Georg Siegbahn (1886–1978) was a Swedish physicist who was awarded the 1924 Nobel Prize in Physics for his discovery of the spectrum of X-rays.

Carl Ludwig Siegel (1896–1981) was a German number theorist. His main research areas are number theory, indeterminate equations, and celestial mechanics. He was awarded the Wolf Prize in Mathematics in

1978. He was one of the greatest mathematicians of the 20th century. It is often said that Andre Weil ranked Siegel as the best mathematician of the first half of the 20th century. In his anthology, Weil commented that one of the right things to do for mathematicians of his generation was to comment on Siegel's work. In fact, both Weil and Hermann Weyl wrote papers explaining and expanding on some of Siegel's papers. Siegel disliked the modern abstract mathematics advocated by the Bourbaki school. As a result, in a sense, he was more like a 19th-century mathematician.

James Harris Simons (1938–2024) was an American mathematician, investor, and philanthropist. He taught at the Massachusetts Institute of Technology, Harvard University, and the State University of New York at Stony Brook. The Chern-Simons form is named after Chern Shiing-Shen and his surname. He received the Veblen Prize from the American Mathematical Society in 1976. In 1982, he switched to the investment industry and founded a highly successful hedge fund. He was one of the major donors to the National Institute of Mathematical Sciences. Simons is a close friend of Professor Yang Chen-Ning. There is a Chern-Simmons Building that provides accommodation for visitors to the Institute of Advanced Studies at Tsinghua University.

Isadore Manuel Singer (1924–2021) was an American mathematician and had been a professor of mathematics at MIT for a long time. He was best known for his work with British mathematician Michael Francis Atiyah in 1962 on the Atiya-Singer index theorem, which established a new interaction between pure mathematics and theoretical physics. In 1982, while a professor at the University of California, Berkeley, he co-founded the Mathematical Sciences Research Institute (MSRI) with Calvin C. Moore and Chern Shiing-Shen.

Stephen Smale (1930–) is an American mathematician who was awarded the Fields Medal in 1966 and the Wolf Prize in Mathematics in 2007. He became famous for his proof of the Poincare conjecture of five dimensions or more, and later turned to the study of dynamical systems with important achievements. When he was a graduate student at the University of Michigan, he was nearly kicked out of the graduate program because of poor grades.

Charles Percy Snow (1905–1980) was an English scientist and novelist. His works *The Masters* and *The New Men* won the 1954 James Tait Black Memorial Prize. He was a friend of mathematician G. H. Hardy and wrote a preface to Hardy's *A Mathematician's Apology.*

Elias Stein (1931–2018) was an American mathematician, member of the National Academy of Sciences, and one of the leading figures in harmonic analysis. Many prominent mathematicians, such as Charles Fefferman and Terence Chi-Shen Tao, were among his students. He was awarded the Steele Prizes in 1984 and 2002, the Rolf Schock Prize in 1993, the Wolf Prize in Mathematics in 1999, and the US National Medal of Science in 2001. He was awarded the Bergman Prize in 2005 for his contributions to the fields of real analysis, complex analysis, and harmonic analysis.

T

Yutaka Taniyama (1927–1958) was a Japanese mathematician who graduated from the University of Tokyo in 1953. As an undergraduate, he read Claude Chevalley's *Theory of Lie Groups*, André Weil's *Foundations of Algebraic Geometry*, and Weil's two other books on algebraic curves and Abelian varieties. These sparked his interest in number theory. The Taniyama-Shimura conjecture, proposed by Taniyama Yutaka and Gorō Shimura, was an important step in British mathematician Andrew John Wiles' eventual solution to Fermat's Last Theorem. Taniyama committed suicide when he was young, and a month later, his fiancée also committed suicide, leaving a note to the world: "We promised each other that no matter where we are, we will never be apart. Now that he is gone, I will follow him."

Richard Taylor (1962–) is a British mathematician working in the field of number theory. He received the 2002 Cole Prize, the 2007 Shaw Prize with Robert Langlands, and the 2015 Breakthrough Prize in Mathematics.

Edward Teller (1908–2003) was a Hungarian-born American theoretical physicist known as the father of the hydrogen bomb. In addition to the hydrogen bomb, he made significant contributions to several fields of physics. He was an early member of the Manhattan Project, worked on the development of the first atomic bomb, and was keen to promote the development of the earliest nuclear fusion weapons (hydrogen bombs). He was one of the founders of Lawrence Livermore National Laboratory, where he served for many years as assistant director and director, and in 1959, presided over the establishment of the Berkeley Space Science Laboratory. Teller is Yang Chen-Ning's thesis tutor.

René Thom (1923–2002) was a French mathematician, best known for his work in establishing the catastrophe theory in 1968. He was awarded the Fields Medal in 1958 for his previous work in differential topology, particularly cobordism theory. He and Grothendieck were both founders of the Institute for Advanced Studies in Paris. The influence and popularity of Grothendieck's seminar at the Institute for Advanced Study put too much pressure on him, and he decided to work on topics that would attract more public attention. Instead, he worked on the catastrophe theory and wrote books such as *Structural Stability and Morphogenesis*.

Shinichiro Tomonaga (1906–1979) was a Japanese theoretical physicist. Engaged in research on quantum electrodynamics and quantum field theory, he proposed the super-many-time theory of quantum field theory in 1941. In 1965, he won the Nobel Prize in Physics together with Julian Schwinger and Richard Feynman.

U

George Uhlenbeck (1900–1988) was a Dutch-born American theoretical physicist. In mid-September 1925, in collaboration with Samuel Goudsmit, he discovered the spin of the electron. Between 1947 and 1966, he was nominated several times for the Nobel Prize in Physics. Wang Chengshu, a female physicist in modern China, was one of his students.

Karen Uhlenbeck (1942–) an American mathematician who won the Abel Prize in 2019 for her pioneering work in geometric partial differential equations, gauge field theory, and integrable systems, as well as for her work in mathematical analysis, geometry, and mathematical physics. She is the first woman to receive the prize. Her best job is related to the mathematical properties of Yang–Mills theories.

Stanislaw Ulam (1909–1984) was a Polish mathematician and nuclear physicist. He worked on the Manhattan Project and, with Edward Teller, invented the Teller-Ulam design of the hydrogen bomb. He also worked on the nuclear-powered space shuttle. In pure mathematics, he worked in ergodic theory, number theory, set theory, and algebraic topology. Ulam had an unusual life, which is documented in his autobiography, *Adventures of a Mathematician*, which is a very interesting book.

Samuel Ullman (1840–1924) was a businessman, poet, and philanthropist. Born in Germany in 1840, he moved to the United States with his family when he was 11. Ullman's poems cover such diverse topics as love, nature, religion, family, lifestyle of friends, etc.

V

Martinus J. G. Veltman (1931–2021) was a Dutch theoretical physicist and professor emeritus at the University of Michigan. He and Gerardus't Hooft won the 1999 Nobel Prize in Physics for their work on renormalization of quantum gauge field theory.

Ivan Matveyevich Vinogradov (1891–1983) was a Soviet mathematician who specialized in analytic number theory. In 1937, he directly proved, without the involvement of the generalized Riemann conjecture, that sufficiently large, odd numbers can be represented as the sum of three prime numbers, also known as Vinogradov's theorem, which was a major step forward in solving Goldbach's conjecture.

Vladimir Voevodsky (1966–2017) was a Russian mathematician who won the Fields Medal in 2002. He developed the new theory of cohomology of algebraic varieties, one of the most remarkable advances in algebraic geometry in the last few decades.

W

Edward Waring (1736–1798) was an English mathematician who formulated what came to be known as the Waring's problem in his book *Meditationes Algebraicae*. Waring speculated that any positive integer could be expressed as the sum of four squares, the sum of nine cubes, the sum of nineteen fourth powers, etc. Based on these speculations, the question was raised: For any given positive integer k, was there such an integer S(k) that every positive integer could be expressed as the sum of S(k) k-powers?

James Dewey Watson (1928–) was an American molecular biologist and one of the leading figures in molecular biology in the 20th century. He and his colleague Francis Crick discovered the double helix structure of DNA, for which they, together with Maurice Hugh Frederick Wilkins were awarded the 1962 Nobel Prize in Physiology or Medicine.

Karl Weierstrass (1815–1897), a German mathematician known as the father of modern analysis, made an important contribution to mathematics by establishing a strict definition of the limits of functions. His story of late success could inspire many who fail to win at the starting line. In fact, he was promoted from a middle school teacher in Berlin to a professor at the age of 40, because he solved the Jacobi inverse problem of a special class of Abelian integrals. The ambition of his life was the establishment of a rigorous theory of Abelian functions. All of his analytical work, familiar to students of mathematics, was prepared for this goal. Unfortunately, he did not achieve this goal in a satisfactory manner. His relationship with his famous schoolgirl, Sofia Kovalevskaya (1850–1891), was also unusual, or rather sad. After Kovalevsky's untimely death, he burned all of her letters.

André Weil (1906–1998) was a French mathematician whose work focused on analytical number theory and algebraic geometry. The Bourbaki Group, a mathematical group he co-founded and led, actively absorbed and developed emerging branches of mathematics, directly influencing the direction of mainstream mathematics development at the time. He was one of the few great mathematicians in the history of mathematics who had a keen interest in studying its history. For example, he came up with the very influential Weil conjecture from reading a paper by Gauss. When he was a college student, he realized the importance of reading the classics of great mathematicians, so he read the anthologies of Riemann. He had high standards and strong opinions about many things. His anthologies contained a great deal of commentary and were of great value to students and mathematicians alike. He was ambitious in his desire to learn all the branches of mathematics, and indeed made profound contributions to some aspects of mathematics. In addition to his in-depth research in algebraic geometry and number theory, some of his ideas led to the famous Chern-Weil theory. Although he was well known in mathematics, his sister Simone Weil (1909–1943) was much better known in the public sphere. The brother and sister's Paris apartment has a plaque commemorating Simone, not André himself.

Simone Weil (1909–1943), sister of André Weil, was a French Jew, mystic, religious thinker, philosopher, and social activist who profoundly influenced post-war European thought. Sylvie Weil, a daughter of André Weil, wrote a moving memoir about the siblings, *At Home with André and Simone Weil.*

Steven Weinberg (1933–2021) was an American physicist. Six years after the discovery of neuter current, or discovery of the Z boson, Weinberg, Sheldon Lee Glashow and Abdus Salam won the 1979 Nobel Prize in Physics together for independently developing unified electro-weak theory based on a spontaneous symmetry breaking mechanism.

Hermann Weyl (1885–1955) was a German mathematician, physicist, and philosopher. He was one of the most influential mathematicians of the 20th century and an important early member of the Institute for Advanced Study in Princeton. He was one of the first to combine general relativity with electromagnetic theory and was the originator of the Abelian gauge field theory. He grew up in the countryside, but when he went to the University of Gottingen, he won the heart of the most beautiful girl there, much to the surprise and dismay of many, especially those from wealthy families in big cities, such as Theodore von Kármán, Qian Xuesen's mentor. While he was still a student, he solved a problem posed by the physicist Lorentz that Hilbert had predicted would be unsolvable in his lifetime.

Hassler Whitney (1907–1989) was an American mathematician. Elected to the National Academy of Sciences in 1945, he was vice president of the American Mathematical Society from 1948 to 1950. He was awarded the National Medal of Science in 1976 and the Wolf Prize in Mathematics in 1982. He was one of the main founders of differential topology and is the author of the *Geometric Integration Theory*. Whitney was also an accomplished mountain climber, and a steep ridge on Mount Cannon is named after him, the Whitney-Gilman Ridge.

Norbert Wiener (1894–1964) was an American applied mathematician who made significant contributions to electronic engineering. He was a pioneer in stochastic processes and noisy signal processing, and he coined the term "cybernetics." He also made great contributions to mathematics, such as the mathematical theory of Brownian motion, the Wiener's Tauberian theorem, Paley–Wiener theorem in harmonic analysis, etc.

Eugene Paul Wigner (1902–1995), born in Hungary, was an American theoretical physicist and mathematician. He laid the theoretical foundation of symmetry in quantum mechanics, made important contributions to the study of the structure of the atomic nucleus, and performed many important works in pure mathematics. In 1963, Wigner, along with Maria Goeppert-Mayer and Johannes Hans D. Jensen, shared the Nobel Prize in Physics for their theoretical contributions to the physics of atomic nuclei and elementary particles, especially the discovery and application of the principle of fundamental symmetry.

Andrew John Wiles (1953–) is a British mathematician who received his Ph.D. at the University of Cambridge in 1977. His research focuses on number theory and related fields, and his greatest contribution is the first complete proof of Fermat's Last Theorem in the 1990s. He was elected a Fellow of the Royal Society in 1984 and a foreign member of the National Academy of Sciences in 1996. After proving Fermat's Last Theorem, he was awarded the 1995–1996 Wolf Prize in Mathematics, the 1996 Prize in Mathematics of the National Academy of Sciences, the Ostrovsky Prize of Switzerland, the Schock Prize (Prize in mathematics) of the Royal Swedish Academy of Sciences, the Fermat Prize of France, the Kohl Prize of the American Mathematical Society in 1997, and the 1998 prize Special Award of the International Congress of Mathematicians, and Shaw Prize in 2005. The Mathematics Building at the University of Oxford is now named after Wiles.

Edward Witten (1951–) is an American mathematical physicist and a professor at the Princeton Institute for Advanced Study. He is a leading expert in string theory and quantum field theory and is the founder of M-theory. Edward Witten is regarded as one of the greatest physicists of our time, and some of his peers even consider him one of Einstein's successors. The International Mathematical Union awarded him the Fields Medal in 1990, and he remains the only physicist to have received this honor.

Y

Hideki Yukawa (1907–1981) was a Japanese theoretical physicist. Yukawa studied the strong interactions inside atomic nucleus that bind protons to neutrons, and in 1935 speculated that mesons should exist between protons and neutrons. In 1947, the British physicist Cecil Frank Powell (1903–1969) discovered a π meson from cosmic rays, which also proved Yukawa's theory. Yukawa became the first Japanese to win a Nobel Prize in 1949.